Stars and Supernovas

Stars and Supernovas

Iain Nicolson

First published in 2001 by
BBC Worldwide Ltd,
Woodlands, 80 Wood Lane,
London W12 0TT

DK PUBLISHING, INC.
www.dk.com

Publisher: Sean Moore
Art Director: Dirk Kaufman
Editorial Director: Chuck Wills

First American Edition, 2001

00 01 02 03 04 05 10 9 8 7 6 5 4 3 2 1

Published in the United States by
DK Publishing, Inc.
95 Madison Avenue
New York, New York 10016

ISBN 0-7894-8160-X

Produced for BBC Worldwide by
Toucan Books Ltd, London

Cover photograph: Science Photo Library

Printed and bound in France by
Imprimerie Pollina s.a. n° L83861
Color separation by Imprimerie
Pollina s.a.

Picture credits:
Page 2 SPL/Celestial Image Co. 6 SPL/Frank Zullo. 9 Galaxy Picture Library, TR; Julian Baker, B. 10 SPL/Pekka Parviainen. 11 SPL/Frank Zullo. 12 Galaxy Picture Library. 13 SPL/Gordon Garradd. 14 Julian Baker. 15 Galaxy Picture Library. 16 The Bridgeman Art Library/O'Shea Gallery, London. 17 AKG London/Erich Lessing. 18 SPL/John Chumack. 19 Julian Baker. 20 SPL/Simon Fraser, TL; Galaxy Picture Library, B. 21 The Bridgeman Art Library/Private Collection. 22-3 SPL/Luke Dodd. 23 SPL/Adam Hart-Davis, BR. 24 SPL/John Sanford, CL; Galaxy Picture Library, R. 25 Galaxy Picture Library, TR, BL. 26 The Art Archive/Bibliothèque Nationale, Paris. 27 SPL/John Sanford, TL; SPL/Luke Dodd, BL. 28 SPL/Celestial Image Picture Co. 31 Roger-Viollet. 32 SPL/John Sanford, TL; Julian Baker, TR. 33 SPL/STScI/NASA. 34 Hubble Heritage Team (AURA/STScI/NASA). 35 Galaxy Picture Library, TL; SPL/David Parker, BR. 36 Anglo-Australian Observatory/Photograph by David Malin. 37 Julian Baker, TR; SPL/Luke Dodd, BL. 38-9 Anglo-Australian Observatory/Photograph by David Malin. 40 STScI/NASA. 41 STScI/NASA, L, R. 42 Julian Baker, TR; SPL, CB; Galaxy Picture Library, BR. 43 Galaxy Picture Library. 44-5 STScI/NASA. 46 SPL/NOAO. 47 SPL/ESA. 48 SPL/ESA. 49 SPL/NOAO, TL; SPL, BR. 50 Anglo-Australian Observatory/Photograph by David Malin. 53 STScI/NASA. 54-5 Anglo-Australian Observatory/Photograph by David Malin. 56-7 Anglo-Australian Observatory/Photograph by David Malin. 58 AIP Emilio Segre Visual Archives, Physics Today Collection. 59 European Southern Observatory. 60-1 Anglo-Australian Observatory/Photograph by David Malin. 63 European Southern Observatory. 64 SPL/NASA. 65 SPL/Frank Zullo, BL; Anglo-Australian Observatory, BR. 66 Anglo-Australian Observatory/Photograph by David Malin. 67 Julian Baker. 68 STScI/NASA. 69 STScI/NASA. 70 Anglo-Australian Observatory/Photograph by David Malin. 71 Anglo-Australian Observatory/Photograph by David Malin. 72 Anglo-Australian Observatory/Photograph by David Malin. 75 Julian Baker. 76-7 Anglo-Australian Observatory/Photograph by David Malin. 78-9 NASA and Chandra Science Center. 80 Anglo-Australian Observatory/Photograph by David Malin. 81 European Southern Observatory. 82 CXC, SAO, NASA, BL; European Southern Observatory, BR. 83 NASA and Chandra Science Center. 85 Julian Baker. 86 Hubble Heritage Team (AURA/STScI/NASA). 87 Julian Baker, TL; Corbis/Bettmann, BR. 88 SPL/STScI/NASA. 89 SPL/STScI/NASA. 90 STScI/NASA. 91 Caltech GRB Team/STScI/NASA. 92-3 European Southern Observatory.

Contents

1 **SEEING STARS** 6

2 **STARS OF MANY KINDS** 28

3 **LIFE CYCLES OF STARS** 50

4 **EXPLODING STARS AND REMNANTS** 72

FURTHER INFORMATION 94

INDEX 94

1

SEEING STARS

1 SEEING STARS

On a clear dark night, when the Moon is not out, the sky seems to be filled with stars, some strikingly brilliant, others barely visible. As the night progresses, the stars move silently across the sky from east to west. They vanish at sunrise – hidden by the glare of the Sun – but reappear in the evening, after sunset. To the ancient sky-watchers the stars seemed permanent and unchanging, always maintaining the same fixed patterns. Although different stars were visible at different seasons of the year, the same star patterns returned to the night sky at the same times of year with unfailing regularity. However, there were five star-like points of light that changed position night by night and week by week, relative to the background stars. These 'wandering stars' were the planets. Although the early sky-watchers were familiar with the stars and planets visible to the naked eye, they had no idea what these points of light really were.

Previous page: This image of the constellation of Taurus (the Bull) captures the haunting beauty of the starlit sky. The V-shaped pattern of stars to the right of centre is the Hyades star cluster.

STARS AND PLANETS: OUR PLACE IN THE UNIVERSE

Nowadays, we know that we live on the surface of planet Earth, a small rocky world that travels around the Sun, a luminous globe of hot gas that is powered by nuclear reactions. The stars are themselves suns, shining in the same sort of way. They look like tiny points of light only because they are so much farther away than our Sun.

Our nearest neighbour in space is the Moon, a rocky body with about a quarter of the Earth's diameter, which travels round the Earth in a period of 27.3 days. It lies at a mean distance of 384,400 km (238,866 miles) – equivalent to travelling 10 times round the equator of the Earth. The Moon emits no light of its own. It shines because it is reflecting sunlight. Located at a mean distance of 149,600,000 km (92,961,440 miles) from the Earth, the Sun is nearly 400 times farther away than the

1. The Sun, seen here as it descends towards the western horizon, is the star around which the Earth revolves.

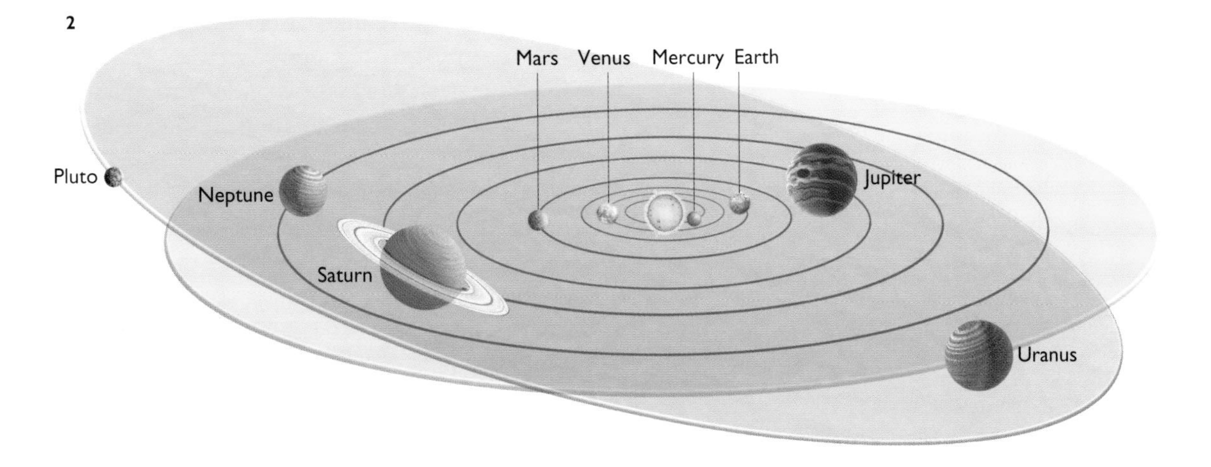

2. The nine planets of our Solar System travel in elliptical orbits around the Sun. The orbit of Pluto, the outermost planet, tilts at a 17-degree angle.

1. The planet Venus, seen here in the dawn sky, appears about 15 times brighter than the brightest star. The Pleiades are centre left.

2. The Milky Way, seen here where it passes through Scorpius and Sagittarius, contains clouds of stars and dark patches of dust.

If you could drive a car through space at 100 km/h, (62 mph) it would take five and a half months to reach the Moon, 170 years to reach the Sun and 46 million years to reach Proxima Centauri (the nearest star).

Moon. Its diameter of 1,400,000 km (869,960 miles) is more than 100 times greater than that of the Earth, and its volume more than a million times greater; a million Earths would fit inside the Sun with room to spare.

The Earth is one of nine planets that travel around the Sun. Mercury and Venus are closer to the Sun than is the Earth, whereas Mars, Jupiter, Saturn, Uranus, Neptune and Pluto are farther away. Like the Moon, the planets emit no light of their own: they shine by reflecting sunlight. Five of them – Mercury, Venus, Mars, Jupiter and Saturn – are bright enough to be seen with the naked eye; Venus is the brightest object in the sky apart from the Sun and Moon.

Stars and galaxies

The nearest star, apart from the Sun itself, is Proxima Centauri, a dim red star too faint to be seen without a telescope, which is some 40 million million km (25 million million miles) away – about a quarter of a million times farther away than the Sun. One way to get a feel for the vast distances that separate the stars is to think of how long it would take a ray of light to travel across them. Travelling at 300,000 km (186,000 miles) per second, light is the fastest-moving thing in the Universe. A ray of light takes just 1.3 seconds to travel from the Moon to the Earth, 8.3 minutes to travel from the Sun, 5.5 hours from Pluto, and 4.2 years from Proxima Centauri. The distance that light travels in one year – nearly 10 million million km – is called a light year. Because its light takes 4.2 years to reach us, the distance of Proxima Centauri is 4.2 light years.

The Sun, together with all the naked-eye stars, is part of a huge star system, or galaxy, which contains at least 100 billion stars. Our galaxy consists of a central bulge, within which most of its stars are concentrated, surrounded by a flattened disc of stars and gas clouds. Within the disc, most of the gas clouds and many of the stars are bunched into a pattern of spiral 'arms'. Our galaxy is about 100,000 light years in diameter. The Sun is about 28,000 light years from the centre of our galaxy, just over halfway from the centre to the edge.

The combined light of the millions of stars in the galactic disc gives rise to a narrow band of hazy starlight called the Milky Way, which stretches right across the sky and can be seen with the naked eye on clear, moonless nights. Our galaxy, which is also called the 'Milky Way Galaxy', is just one of billions of similar galaxies in the observable Universe.

2

THE REVOLVING SKY

To the earliest civilizations, it seemed as if the Earth was flat and the sky was a dome suspended above it supported, perhaps, by distant mountain ranges. Such views reflected the environment in which these civilizations existed. With no artificial lighting or haziness created by pollution, the sky really would have looked like a dome on which the stars were set.

Great strides were made by the ancient Greeks. By the beginning of the 5th century BC, a number of Greek philosophers had concluded that the stars were attached to a sphere that rotated around the Earth once a day, and that the Earth itself was also a sphere. By the 4th century BC, Eudoxus had developed a theory whereby a set of concentric spheres carried the Sun, Moon, planets and stars around the Earth. The Earth was believed to lie at the centre of the Universe. Although Aristarchus, in the 3rd century BC, suggested that the Earth and the planets travelled around the Sun and that the daily movements of the Sun and stars were caused by the Earth rotating on its axis, it was not until the 17th century AD that the notion of a central Earth was abandoned.

1

1. This time-lapse series of exposures shows the track of the Sun across the sky in the course of a day.

2. Star trails. During this exposure, of nearly 12 hours' duration, the Earth's rotation has turned the image of each star into a semicircular trail centred on the south celestial pole.

2

Nowadays we know that the stars are themselves suns which lie at vast and very different distances from the Earth. Nevertheless, when trying to describe the positions and motions of the Sun, stars and planets in the sky, it is helpful to imagine that they are attached to a huge sphere, called the 'celestial sphere', which rotates round the Earth. Like the Earth, the celestial sphere has north and south poles and an equator. The apparent rotation of the celestial sphere causes the Sun and stars to move across the sky parallel to the celestial equator.

If you were at the North (or South) Pole of the Earth, the north (or south) celestial pole would be vertically overhead and the celestial equator would coincide with the horizon. Stars would move parallel to the horizon, and would neither rise nor set. At the North Pole, the northern half of the celestial sphere is visible all the time and the

There are about 5800 stars on the entire celestial sphere that are bright enough to be seen, under ideal conditions, with the naked eye. At any particular time, half of them are below the horizon.

southern part is permanently hidden below the horizon. Conversely, from the South Pole, the southern half of the celestial sphere is always visible, and the northern half can never be seen. Viewed from the Earth's Equator, the celestial equator crosses the horizon at right angles, and passes directly overhead, whereas the north and south celestial poles coincide with the north and south points of the horizon. Stars rise vertically at the eastern horizon and set vertically at the western horizon. Only half of the celestial sphere can be seen at any instant as the sphere spins round, but every part of it can be seen at one time or another.

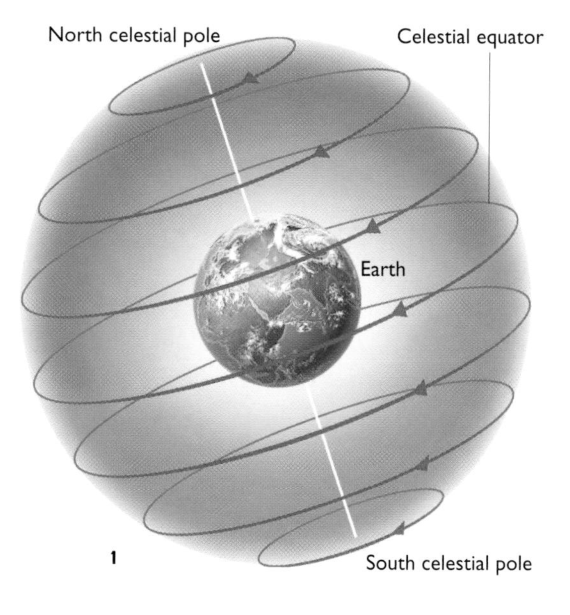

Circumpolar stars

Viewed from anywhere else on our planet, some of the stars remain above the horizon at all times, some never rise at all, and others rise and set at an angle to the horizon. Stars that never set are called circumpolar stars. As the celestial sphere rotates, each circumpolar star traces out a circle around the celestial pole, every part of which is above the horizon. Which stars are circumpolar and which are never seen depends on your latitude (the angle between the Equator and the place at which you are located). Wherever you are on Earth, the altitude of the celestial pole (the vertical angle between the horizon and the celestial pole) is equal to your latitude. So if you live at latitude 51.5° N (the latitude of London), the celestial pole will be 51.5° above your horizon, and any star lying within 51.5° of the south celestial pole will never be seen.

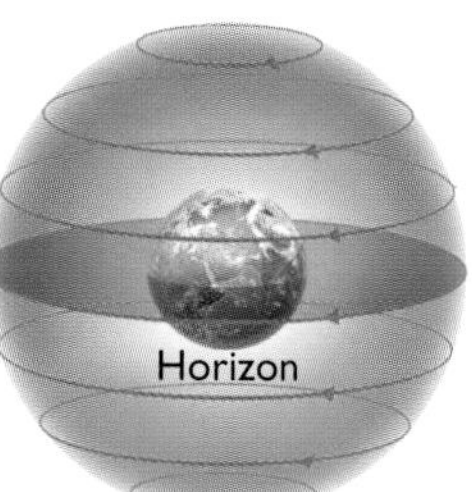

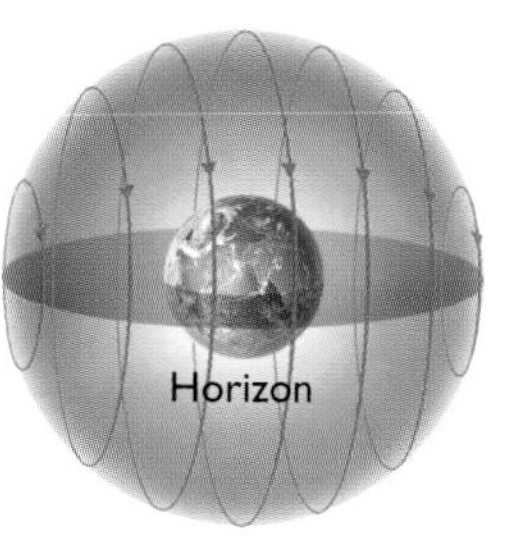

1. Owing to the Earth's rotation, the celestial sphere (above) appears to rotate from east to west, in the direction of the arrows, around an axis that joins the north and south celestial poles. Viewed from either of the Earth's poles (left), stars appear to move parallel to the horizon. Viewed from the Equator (bottom left), stars rise and set vertically.

STAR PATTERNS

Thousands of years ago, sky-watchers began to identify patterns among the stars. These star patterns, or constellations, may originally have been devised as a way of dividing up the night sky into identifiable chunks and as an aid to recognizing which star patterns were visible at important times in the agricultural year. The earliest constellations were named after animals. For example, the Sumerians, who lived in the Middle East more than 5000 years ago, appear to have identified a lion, a bull and a scorpion among the stars.

The ancient Greeks named many of their constellations after creatures and personalities that figured in their myths and legends. For example, Perseus was a mythological Greek hero who cut off the head of Medusa, one of three legendary creatures, called Gorgons, whose gaze could turn all those who looked on her to stone. While returning from his venture, Perseus came across Andromeda, the daughter of King Cepheus and Queen Cassiopeia, who was about to be devoured by a sea monster. He saved Andromeda by showing Medusa's head to the monster and thereby turning it to stone. Andromeda, Cepheus, Cassiopeia and Perseus are all represented by constellations, and Medusa's head is marked by the star Algol.

The most magnificent of the Greek constellations is Orion, which represents a mythological hunter

2. The shapes of various constellations are picked out here by lines joining their principal stars. Clockwise from top left: Pegasus, Equuleus, Delphinus, Aquila, Scutum, Sagittarius, Capricornus, Aquarius.

1. An engraving by the German artist Georg Christoph Eimmart (1638–1705) shows representations of the constellations on two hemispheres of sky. It also shows the motion and phases of the Moon and the planetary movements.

who was killed by a scorpion. Its distinctive features include a line of three bright stars, which represents his 'belt', and two particularly bright stars – Betelgeuse (which marks his right shoulder) and Rigel (which marks his left foot).

The Greek constellations were subsequently given Latin names, such as Ursa Major (the Great Bear), Canis Major (the Large Dog), Leo (the Lion), Taurus (the Bull), Scorpius (the Scorpion), and so on.

Ptolemy of Alexandria, who lived in the 2nd century AD, listed 48 constellations. All of them are still to be found on present-day star maps, except for one huge constellation, Argo Navis (which represents the ship on which the mythological hero Jason set forth in quest of the legendary Golden Fleece), which has been broken into three: Carina (the Keel), Puppis (the Poop) and Vela (the Sail). Since Ptolemy's time, more constellations have

2. A painting (*c.* 1476) by Justus van Gent of the 2nd-century Greek astronomer and mathematician, Ptolemy.

THE STARS THROUGH DIFFERENT EYES

Different cultures identified different star patterns and named them in their own fashion. The constellations that were identified long ago by the Chinese, African and American peoples were markedly different from those devised by the ancient Greeks. A few, however, had strong similarities. For example, the Chinese identified the seven stars of the Plough, or Big Dipper, and called this pattern the Northern Bushel. They were familiar with the three stars of Orion's belt, and called them Shen (the Union of Three). These three stars, though named in many different ways, were well known throughout the African continent. On the opposite side of the Atlantic, the Mayan peoples of Central America included a Scorpion among their constellations.

1

been added, including those that are too far south to have been seen by the ancient Greeks. Nowadays, the entire celestial sphere is divided up into a grand total of 88 constellations.

Individual bright stars have also been named. While some of the names, such as Sirius, Procyon and Castor, are of Greek origin, and a few, such as Regulus and Polaris, are Roman, the great majority were given by Arabian astronomers. Among the many Arabic names are Betelgeuse (in Orion), Algol (in Perseus), Aldebaran (in Taurus) and Deneb (in Cygnus). In 1603, the German astronomer Johann Bayer assigned Greek letters to the brighter stars in each constellation. In principle, the brightest star was denoted by alpha (α), the next brightest by beta (β), and so on, although in practice the labelling was not always in the correct order of brightness. To the letter was added the Latin possessive form of the constellation name. According to Bayer's scheme, Deneb is Alpha Cygni (Alpha of Cygnus), Sirius is Alpha Canis Majoris and Algol is Beta Persei.

Constellations, as such, have no physical significance. Although the stars that make up a constellation happen to lie in similar directions as viewed from the Earth, they are often at very different distances. For example, of the bright stars in Orion, Betelgeuse lies at a distance of about 300 light years, Rigel at about 900 light years and Mintaka (at the northwestern end of the 'belt') at more than 2000 light years. Rigel and Mintaka are farther from Betelgeuse than we are!

There are, however, some clusters of stars that are physically associated with each other. Best known of these is the Pleiades, a tight little cluster of stars located in the constellation of Taurus. Under good conditions, average-sighted observers can see six or seven stars in this cluster, but keen-sighted observers can see up to twice as many. With a telescope or a pair of binoculars, many more can be seen.

SEASONAL CHANGES

If the Earth's axis were perpendicular to the plane of its orbit, the Sun would always be directly overhead at the Equator, and would be on the horizon when viewed from the poles. Day and night would be of equal duration – each 12 hours long – everywhere on the surface of our planet. However, because the Earth's axis is actually tilted from the vertical by an angle of 23.45°, the Earth experiences an annual cycle of seasons – spring, summer, autumn and winter – during which the amount of daylight increases and decreases.

On or around 21 March each year, the Sun is vertically overhead at the Equator, and every point on the Earth's surface experiences 12 hours of daylight and 12 hours of night. This date, known as

1. The Pleiades star cluster contains about 500 stars. The hazy patches surrounding the brightest stars are caused by their light reflecting from clouds of dust-laden gas.

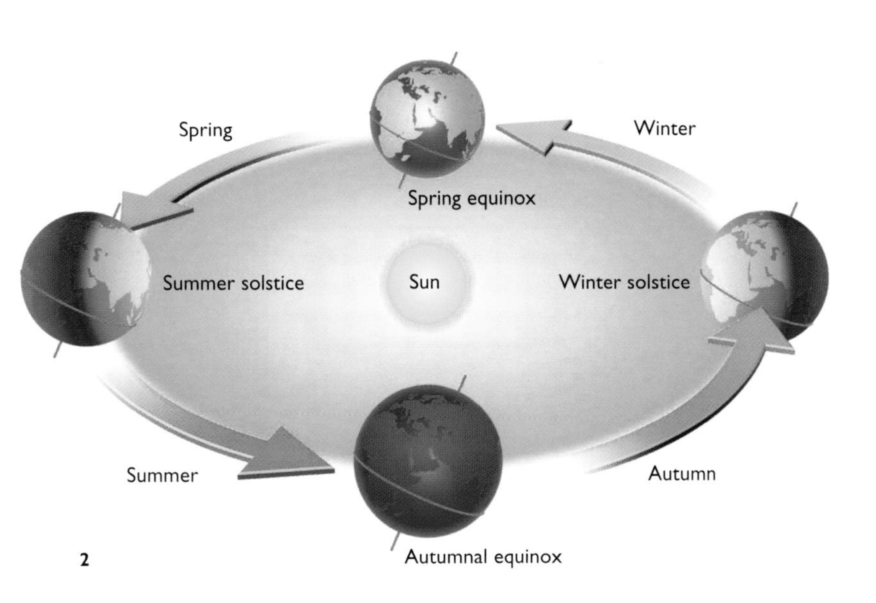

2. The seasons occur because the Earth's tilt, combined with its motion around the Sun, cause periodic variations in the amount of sunlight falling on the north and south hemispheres.

2

1. The midnight Sun over Greenland. In polar regions, during the height of summer, the Sun does not rise or set, but remains above the horizon for 24 hours a day.

2. and 3. Two views of the evening sky, taken facing in the same direction. The winter view (left) includes Orion and Sirius; and the summer view (right) Virgo and Libra.

the vernal equinox, marks the official start of spring in the northern hemisphere. By around 21 June, the Earth has travelled a quarter of the way around its orbit, and the Sun is vertically overhead at latitude 23.45°N (the tropic of Cancer). This is the summer solstice (midsummer's day). Everywhere within 23.45° of the North Pole (i.e. inside the Arctic Circle) experiences continuous daylight, and everywhere within 23.45° of the South Pole (inside the Antarctic Circle) experiences continuous night.

Autumn and winter

On or around around 22 September, the Sun is once again directly over the Equator, and autumn begins. In a further three months – on or around 22 December – it is midwinter's day (the winter solstice). Everywhere inside the Arctic Circle is dark, while everywhere within the Antarctic Circle enjoys continuous daylight. Three months later, the Earth has travelled once around the Sun, and spring returns to the northern hemisphere.

4. A 19th century engraving of the winter sky. The chart shows the celestial equator and the ecliptic, together with representations of various constellations, including two of the zodiacal constellations – Taurus (the Bull) and Aries (the Ram).

4

If we could see the stars in daylight, we would see the Sun in front of the background stars. Day by day, we would be able to see how its position changes, relative to the stars, as the Earth moves around the Sun. In a year, the Earth travels once around the Sun, while the Sun – as seen from the Earth – appears to travel once around the celestial sphere. The path traced out by the Sun is called the 'ecliptic', and the band of stars through which the ecliptic passes is known as the zodiac. In the course of a year, the

To the ancient Egyptians, the Milky Way was the body of the goddess Nut, who gave birth to the Sun god Ra at each winter solstice.

Sun travels through the 12 traditional zodiacal constellations: Aries (the Ram), Taurus (the Bull), Gemini (the Twins), Cancer (the Crab), Leo (the Lion), Virgo (the Virgin), Libra (the Scales), Scorpius (the Scorpion), Sagittarius (the Archer), Capricornus (the Goat), Aquarius (the Water Carrier) and Pisces (the Fish). In addition, between 30 November and 17 December each year it passes through Ophiuchus (the Serpent Bearer).

As the Earth moves around the Sun, its night-time hemisphere (the side that points away from the Sun) faces in a continuously changing direction. Because of this, each star or constellation rises about four minutes earlier each successive evening, and two hours earlier each successive month until, after one complete year, it again rises at the same time of night. Different stars and constellations are seen to best advantage at different times of year. For example, when viewed from the northern hemisphere, Orion is due south, and highest above the horizon, at around midnight in December whereas it is hidden by the Sun's glare in June.

The early civilizations associated particular stars with events in the seasons. For example, in parts of Africa and the Americas, the arrival of the Pleiades star cluster in the dawn sky signalled that now was the best time to plant crops. Among early agricultural societies, observations of the stars were of vital importance.

1. Taurus includes the bright star Aldebaran (left), the V-shaped Hyades (left of centre) and the Pleiades (top right).

SIRIUS AND THE CALENDAR YEAR

The fertile land adjacent to the River Nile (below) is of vital importance to Egypt. Five thousand years ago, the Egyptians already knew that there were 365 days in a year. Around 2500 BC they realized that the date of the annual Nile flood, which irrigated the land, coincided with the time of year at which the star Sirius (which they called Sothis) first became visible in the eastern sky just before sunrise (this event is called the 'heliacal' rising of Sirius). As time went by, they realized that the date of the heliacal rising, and with it the date of the flood, was gradually slipping through their calendar year at a rate of one day every four years, and that after 1460 years (365 x 4) it would return to the same date in their calendar. From this they deduced that the year consists not of a whole number of days, but of 365 1/4 days. In modern times, we keep the calendar year in step with the seasons by adding an extra day to the month of February every four years.

SIGNPOSTS IN THE SKY

Although at first glance the night sky seems a confusing jumble of stars, finding your way around becomes much easier once you have identified a few key constellations. These can then be used as signposts to help you identify others.

Best known of the northern hemisphere constellations is Ursa Major (the Great Bear). Its seven brightest stars make up a distinctive shape (or 'asterism') that looks like a saucepan with a bent handle, and which is known as the Plough (in the UK) or the Big Dipper (in the USA). The Plough is circumpolar from any place north of latitude 41°N, for example, the whole of the British Isles and the northern parts of the USA.

The two stars that make up the side of the saucepan opposite to the handle are Dubhe and

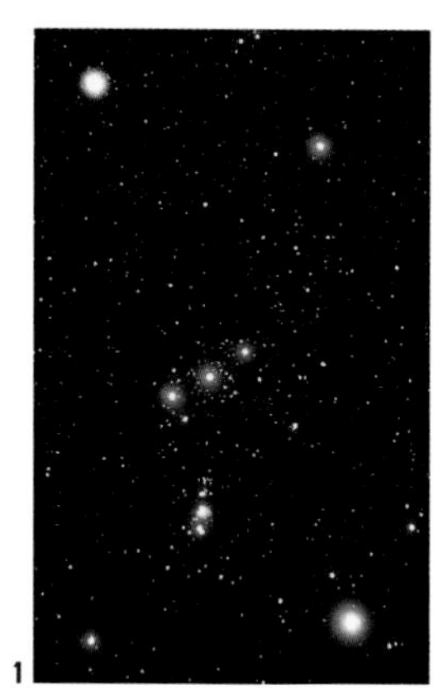

1. The constellation of Orion, the Hunter. The bright star at the top left is Betelgeuse, and the bright star at the lower right is Rigel. In the centre is a line of three stars, Alnilam, Alnitak and Mintaka, which represents Orion's belt.

THE NORTHERN HEMISPHERE

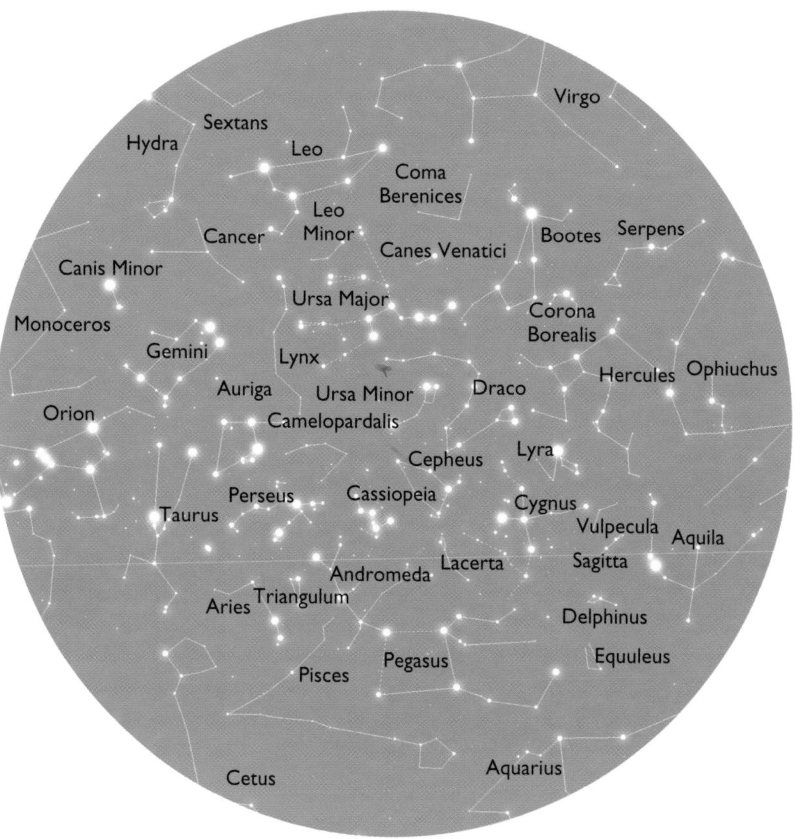

Merak. They are known as the Pointers because a line from Merak through Dubhe leads towards Polaris, a moderately bright star, which, because it lies within 1 degree of the north celestial pole, is known as the Pole Star (or North Star). Polaris lies about 30 degrees away from Dubhe, at the tail of Ursa Minor (the Little Bear). The middle star of the handle is Mizar. A line from Mizar through Polaris, and extended for a further 30 degrees, leads to the

2

2. The lines on this image link together the seven stars that form the Plough, a conspicuous pattern that lies within the constellation of Ursa Major. The two stars on the right of the Plough are Merak (lower) and Dubhe.

THE SOUTHERN HEMISPHERE

W-shaped constellation of Cassiopeia. If the curve of the handle is continued beyond its end, for about 30 degrees, it leads to Arcturus, the brightest star in Boötes (the Herdsman).

Orion, a striking pattern of stars visible from any part of the Earth, is another key constellation. Viewed from the northern hemisphere, Orion dominates the southern part of the sky during the winter. Three bright stars at the centre of the constellation form a line, known as Orion's belt. The line of Orion's belt leads southeastwards, towards Sirius, the brightest star in the sky and the principal star in Canis Major (the Large Dog). Following the line of the belt upwards to the northwest leads to orange-red Aldebaran, the brightest star in Taurus (the Bull). Continue a little farther in the same direction and you will arrive at the Pleiades. Other nearby stars and constellations include Procyon (the brightest star in Canis Minor), Castor and Pollux (the brightest stars in Gemini) and Capella (the brightest star in Auriga).

A conspicuous feature of the summer skies is a pattern of three brilliant stars called the Summer Triangle. This is not a constellation in its own right, but consists of the brightest stars from three constellations: Vega in Lyra (the Lyre), Deneb in Cygnus (the Swan) and Altair in Aquila (the Eagle). Seen from the northern hemisphere, the Summer

PRECESSION AND THE STARS

When a spinning top is tilted from the vertical, it begins to wobble. The Earth behaves in a similar way, its axis sweeping slowly round in a conical pattern over a period of 25,800 years. During this time, the positions of the north and south celestial poles trace out circles in the sky and the position of the vernal equinox (the point at which the Sun crosses the celestial equator around 21 March each year) migrates through the constellations of the zodiac. This phenomenon is called 'precession'. At present, the north celestial pole lies close to Polaris, but 4500 years ago (when the great Egyptian pyramids were being built) it lay close to the star Thuban in the constellation of Draco (the Dragon). In about 12,000 years' time, it will be close to the bright star Vega. Two thousand years ago, astrologers divided the zodiac into 12 equal-sized 'signs' (right), which had the same names as the zodiacal constellations. At that time, the vernal equinox lay in Aries. Because of precession, it is now in Pisces. The astrological signs of the zodiac no longer coincide with the constellations that bear the same name.

1. Sirius, the brightest star in the sky, lies in the constellation of Canis Major.

2. The kite-shaped pattern of stars at the centre of this image is the Southern Cross. The bright stars at the lower left are Alpha and Beta Centauri.

Triangle is high in the southern sky at around midnight in July, 10 p.m. in August and 8 p.m. in September (add 1 hour for summer time).

The southern stars

Centaurus (the Centaur) is a key constellation of the far southern sky. Its brightest star, Alpha Centauri (also known as Rigil Kent), is the third brightest star in the sky and is the nearest star to Earth, other than the Sun itself, that is visible to the unaided eye. A line from Alpha through Beta (Agena) leads towards Crux Australis (the Southern Cross) a compact but very distinctive constellation, which lies about 30 degrees from the south celestial pole. Although a line from Gamma Crucis (Gamma in Crux) through Alpha Crucis leads close to the south celestial pole, there is no conspicuous star anywhere near the south pole of the sky. Seen from the latitude of Australia, for example, the Southern Cross is at its highest at around 1 a.m. in March, 11 p.m. in April, 9 p.m. in May and 7 p.m. in June.

Canopus, the second-brightest star in the sky, lies in the constellation of Carina and is highest in the sky, when viewed from Australia, at about 9 p.m. in February. Achernar lies at the southern end of the long, inconspicuous and straggly constellation of Eridanus (the River) and is the only bright star in that region of the sky. It is highest in the sky at around 9 p.m. in late November. Crux, Alpha Centauri, Canopus and Achernar all provide useful markers in the far southern sky. When these key stars and constellations have been identified, finding the others is then a question of filling in the gaps.

2

STARS OF MANY KINDS

2 STARS OF MANY KINDS

Stars are luminous globes of gas, similar to our Sun, which generate energy by means of nuclear reactions that take place deep inside their cores. Compared to the Sun, some are larger, some are smaller, some are hotter, some are cooler, some are far more luminous and others are very much dimmer. All of them are so far away that, barring a few exceptional cases, no telescope yet built will show their surfaces, so they look like tiny discs. Although binoculars or telescopes will make the stars seem much brighter, and will reveal many more stars than can be seen with the naked eye, those stars still look like points of light. Most of what we know about the stars has been learned through careful measurements of their brightnesses, positions and colours, by looking in detail at the light and other radiations that they emit and by analysing any changes that take place in any of their measurable properties.

Previous page: Mintaka (top right) and Alnilam are, respectively, the most westerly and central stars of Orion's belt. Both are about 10,000 times more luminous than the Sun.

DISTANCE AND LUMINOSITY

During the 2nd century BC, the Greek astronomer Hipparchus divided the stars into six classes, or magnitudes, with the brightest stars in the sky being of first magnitude and the faintest visible stars of sixth magnitude. Some 2000 years later, in 1856, the English astronomer N.R. Pogson put the magnitude scale on a firm mathematical footing. On this scale, a difference of five magnitudes corresponds to a factor of 100 in brightness, so, for example, a star of magnitude 1 is 100 times brighter than a star of magnitude 6.

Stars that are fainter than naked-eye visibility have higher values of magnitude. For example, a star that is 100 times fainter than the naked eye limit would be of magnitude 11 (five magnitudes fainter than a star of magnitude 6). Objects that are brighter than first magnitude have fractional, zero, or minus values of magnitude. For example, the apparent magnitude of Vega is 0.03, and the apparent magnitude of Sirius (the brightest star in the sky) is -1.46. When the planet Venus is at its brightest, its apparent magnitude is -4.4.

Astronomers measure the distances of stars by using the principle of 'parallax'. To see how this works, try the following experiment. Hold up one finger at arm's length, close your left eye, and then line up your finger with a distant object, such as a tree. Now, without moving your finger, close your right eye and open your left. You will find that your finger is no longer lined up with the distant object.

1. A fanciful engraving of the Greek astronomer, Hipparchus, observing from Alexandria in the 2nd century BC. The telescope was not invented until the 17th century AD.

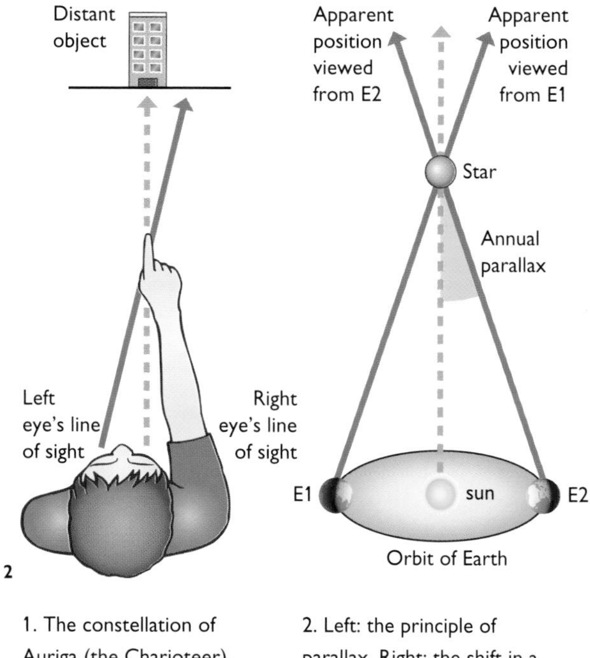

1. The constellation of Auriga (the Charioteer). The brightest star is Capella, a yellow star which lies 43 light years from Earth.

2. Left: the principle of parallax. Right: the shift in a star's position when viewed from opposite sides of the Earth's orbit is used to calculate its distance.

This occurs because your eyes are separated by a few centimetres and each eye, therefore, is looking along a slightly different direction towards the foreground object (your finger). The observed shift in position is called parallax. In order to measure the distance of a nearby star, astronomers measure its position when the Earth is on one side of the Sun (say, in January), and again when the Earth is on the opposite side of the Sun (say, in July). As the measurements have been made from two widely separated points, the observed positions will be slightly different. The maximum amount by which a star shifts from its mean position in the sky during the course of a year is called 'annual parallax'. If the parallax can be measured and the diameter of the Earth's orbit is known (300,000,000 km/186,000,000 miles), simple trigonometry can be used to calculate its distance.

Stars are so far away that their parallaxes correspond to very tiny angles indeed. Angles are usually expressed in degrees, each degree being subdivided into 60 'minutes of arc' and each minute of arc being subdivided into 60 seconds of arc (abbreviated to 'arcsec'). One arcsec is equivalent to one-sixtieth of a sixtieth (1/3600) of a degree. The distance at which a star would have an annual parallax of precisely 1 arcsec is called a 'parsec', and is equivalent to 3.26 light years. The greater the distance of a star, the smaller its annual parallax. For example, a star at a distance of 2 parsecs would have

The Pistol Star, so called because of the shape of the glowing gas cloud that surrounds it, is about 100 times as massive as the Sun, and 10 million times more luminous.

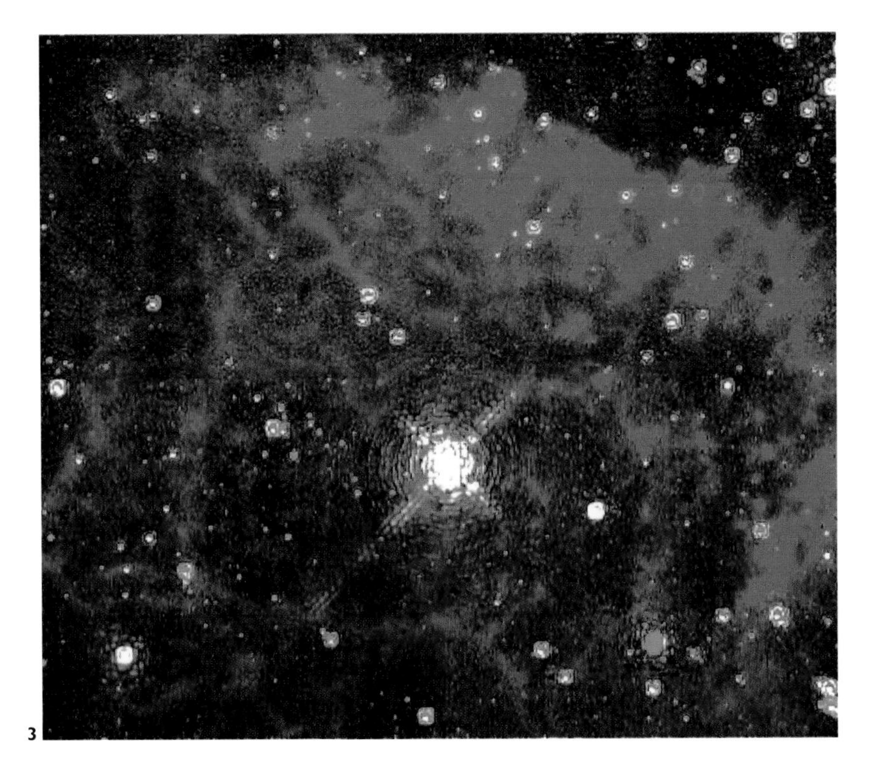

3. The Pistol Star (roughly centre) is one of the most luminous stars in the galaxy. Located close to the centre of the galaxy, it lies at a distance of 25,000 light years from Earth.

an annual parallax of 0.5 arcsec, a star 10 parsecs (32.6 light years) distant would have an annual parallax of 0.1 arcsec, and so on. The parallax of the nearest star (Proxima Centauri) is 0.772 arcsec, equivalent to a distance of 1.3 parsecs (4.2 light years).

We cannot tell from its apparent brightness alone whether a particular star is a relatively dim one that happens to be quite near to us, or a highly luminous one that is a long way away. However, if we know how much light is arriving from the star (by measuring its apparent magnitude), and we know how far away it is (by measuring its parallax), we can calculate its luminosity. The most luminous stars are hundreds of thousands of times more luminous than the Sun; the least luminous shine with only a few hundred thousandths of the Sun's power.

COLOUR AND THE SPECTRUM

Light is a form of electromagnetic wave – an electric and magnetic disturbance that travels through space at a speed of 300,000 km (186,000 miles) per second. It behaves in many ways like a wave on water, the distance between successive wavecrests being the 'wavelength'. Our eyes respond to different wavelengths of light by perceiving different colours: red, orange, yellow, green, blue, indigo and violet, in order of decreasing wavelength. Red light has a wavelength of about 700 nanometres (one nanometre, denoted by nm, is equal to a billionth of a metre), whereas blue light has a wavelength of about 400 nm.

The colour of a hot body is related to its temperature. If a lump of iron is heated in a

1

1. The Sagittarius Star Cloud (SGR-1) at the centre of our galaxy. The hottest stars are blue, the coolest are red.

2. The blue region of the spectrum of the star Altair, which is hotter than the Sun, contains a dark hydrogen line.

3. The solar spectrum, displayed as a series of strips from short wavelengths (top) to long wavelengths.

2

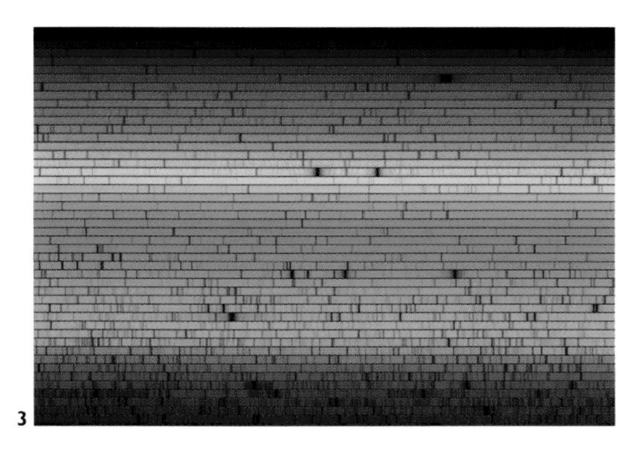

3

furnace, it begins to glow dull red. Gradually, as its temperature increases, it becomes, successively, orange, yellow, then white. Stars behave in a similar sort of way. Red stars are relatively cool, yellow stars are hotter, white stars are hotter still, and the hottest stars of all are blue.

When white light, which is a mixture of all wavelengths, passes through a glass prism, its different wavelengths are bent, or refracted, by differing amounts (red light is refracted least, blue and violet most) and emerge as an unbroken rainbow band of colours which is called a 'continuous spectrum'. A hot, dense, gaseous body emits a continuous spectrum. If that spectrum passes through a low-density gas, atoms in that gas will absorb light at particular wavelengths, thereby imprinting a pattern of dark lines (absorption lines) on the spectrum. The spectrum of a normal star consists of a

SEEING BEYOND THE VISIBLE

Electromagnetic waves extend over a huge range of wavelengths, most of which are invisible to the human eye. The full range (the electromagnetic spectrum) is divided into wavebands. From the shortest to the longest wavelengths, these bands are: gamma ray, X-ray, ultraviolet, visible, infrared, microwave and radio. Although the Sun is brightest at visible wavelengths, it emits all kinds of radiation, from X-rays to radio waves. Extremely hot stars emit mainly ultraviolet radiation, whereas extremely cool stars emit strongly at infrared wavelengths. Although visible light, a few of the infrared range and various microwave and radio wavelengths penetrate to ground level, most incoming wavelengths are absorbed by our atmosphere, or reflected back into space. Gamma rays, X-rays, most of the ultraviolet and infrared must be studied by instruments on satellites, such as IRAS (right), and spacecraft.

Barnard's Star will make its closest approach to the Solar System in the year 11,800, when its distance will be about 3.85 light years.

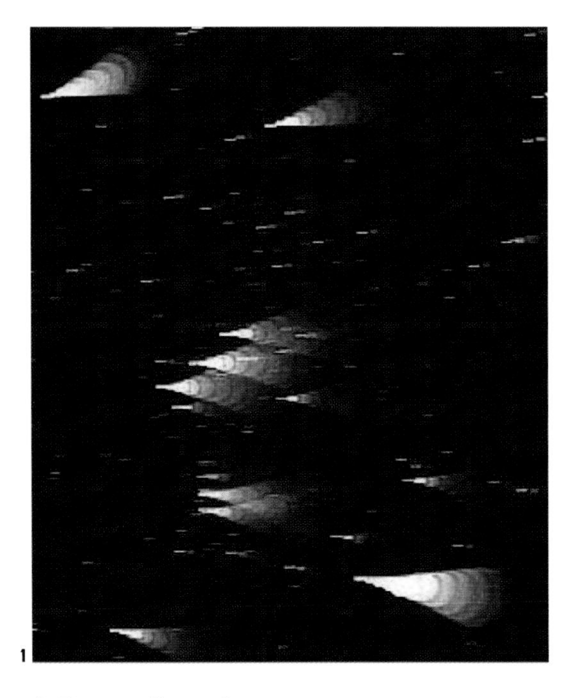

1. The constellation of Orion, with star images trailed out to show colour variations.

continuous spectrum and numerous absorption lines that are produced when its light passes through its cooler outer layers. Because each chemical element produces its own particular pattern of lines, astronomers can find out a star's chemical composition by studying its spectral lines (the lines in its spectrum).

Which lines are prominent depends not only on the chemical composition of the star, but also on its temperature. Stars are divided into classes according to the appearance of their spectra (the plural of spectrum). In order of decreasing temperature, the principal spectral classes are denoted by the letters O, B, A, F, G, K and M. Each class is divided into 10 sub-classes denoted by a number from 0 to 9. The Sun is of type G2. O-type stars have surface temperatures of 30,000°C or more, G-type stars around 6000°C and M-type stars in the region of 3000°C. These stars are bluish, yellow and red respectively .

If a source of light is receding from us, its light waves are stretched, whereas if the source is approaching, its light waves are squeezed together. This phenomenon, called the Doppler effect, also affects the wavelengths of spectral lines. If a star is moving away from us, its spectral lines are shifted towards the long-wave (red) end of the spectrum, whereas if the star is approaching, its spectral lines are shifted towards the short-wave (blue) end of the spectrum. By comparing the observed wavelengths of the lines in the spectrum of a star with the wavelengths those lines would have if the star were stationary, astronomers can measure the speed at which it is approaching or receding.

GIANTS, DWARFS AND THE H-R DIAGRAM

Information about temperatures and 'luminosities' (light outputs) of stars can be displayed on a very useful diagram called the Hertzsprung-Russell (H-R) diagram, devised in the early 20th century by the Danish astronomer Ejnar Hertzsprung and the American astronomer Henry Norris Russell. On a diagram of this kind, luminosity is plotted in the vertical direction, and temperature (or spectral class) is plotted on the horizontal axis. Luminosity increases from bottom to top, on a scale in which the luminosity of the Sun = 1. Because spectral classes are plotted from left to right in the order O, B, A, F, G, K, M, temperature decreases from left to right. Each star is plotted on the diagram as a point corresponding to its luminosity and temperature.

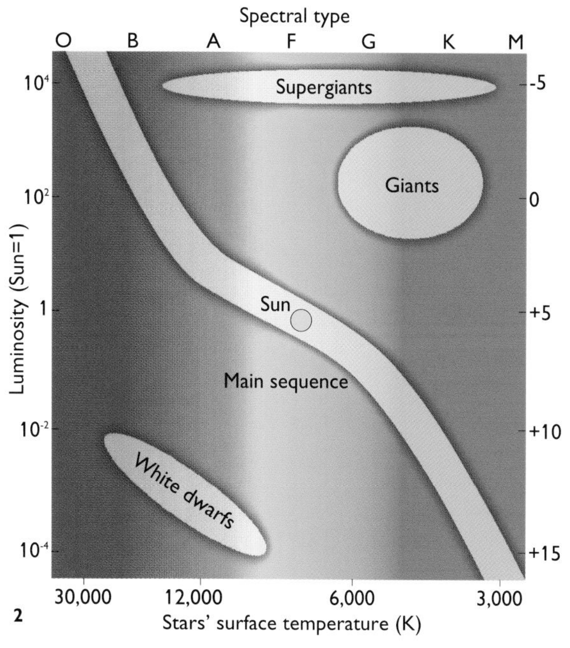

2. The H-R diagram shows the relationship between the colours, temperatures and luminosities of stars. Most stars lie within the main sequence.

3. Antares (left), a red supergiant in the constellation of Scorpius, is seen here surrounded by wispy nebulosity (haziness).

For example, the Sun would be represented by a point corresponding to a temperature of about 6000°C (or spectral type G2) and a luminosity of 1.

When large numbers of stars are plotted on a diagram of this kind, most of them lie in a band, called the main sequence, that slopes downwards from the upper left (high temperature and high luminosity) to the lower right (low temperature and low luminosity). The Sun is a main sequence star.

Giants and dwarfs

Some stars lie above and to the right of the main sequence. Compared to main sequence stars of the same temperature, these stars are much more luminous. If two stars have the same surface temperature, each star emits the same amount of light from each square metre of its surface. If one star is more luminous than the other, it must have a larger surface area (in order to emit a greater total amount of light) and must, therefore, be a bigger star. Most of the stars that are found above and to the right of the main sequence are red giants ('red' because they are cool, and 'giant' because they are much larger than ordinary stars). A typical red giant has a temperature of about 3300°C, a luminosity 100–1000 times that of the Sun and a diameter 20–100 times greater than that of the Sun.

Supergiants, which appear at the top of the diagram, are even more luminous – typically 10,000–100,000 times as luminous as the Sun. Red supergiants are the largest stars of all. Betelgeuse, a red supergiant in the constellation of Orion, is so large that if it were placed at the centre of the Solar

1

1. A bright supergiant lies at the centre of this nebula. The nebulosity is caused by light reflected from particles of dust that have condensed out of material that this star is losing from its distended surface. The four spikes extending from the star are artificial features produced by structures inside the telescope.

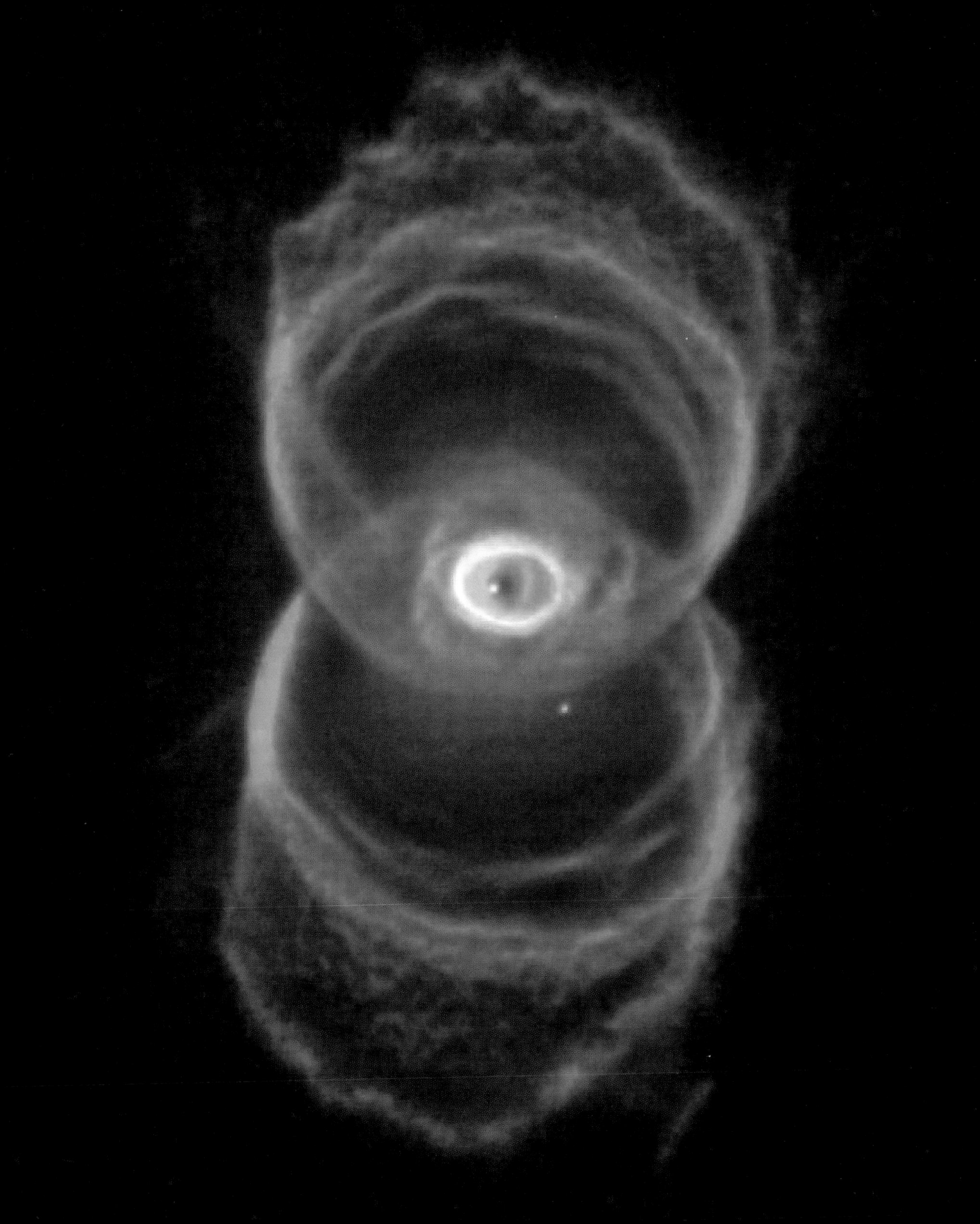

System, it would extend beyond the orbits of the planets Mercury, Venus, the Earth and Mars.

Other stars lie well below and to the left of the main sequence. Since they have high surface temperatures, but very low luminosities, they must be tiny compared to main sequence stars. These stars are called white dwarfs ('white' because of their high surface temperatures, 'dwarfs' because of their small sizes). A typical white dwarf has a radius of between a fiftieth and a hundredth of the Sun's radius, and is similar in size to the planet Earth.

A typical white dwarf contains nearly as much material as the Sun, squeezed into a volume that is a hundred thousand to a million times smaller than that of the Sun. The mean density of a white dwarf is so high that a sugar-lump size piece of white dwarf material, if brought back to the Earth, would weigh about a tonne. By contrast, matter is spread so tenuously through the body of a typical red giant that, averaged out, it has less than one-hundredth of the density of the Earth's atmosphere at sea level.

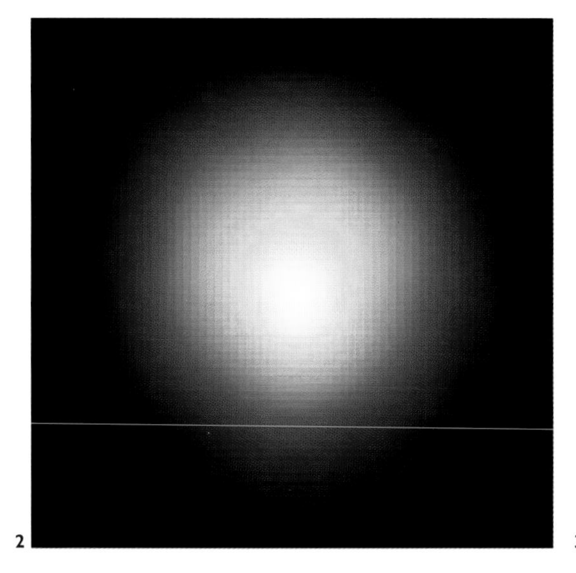

2

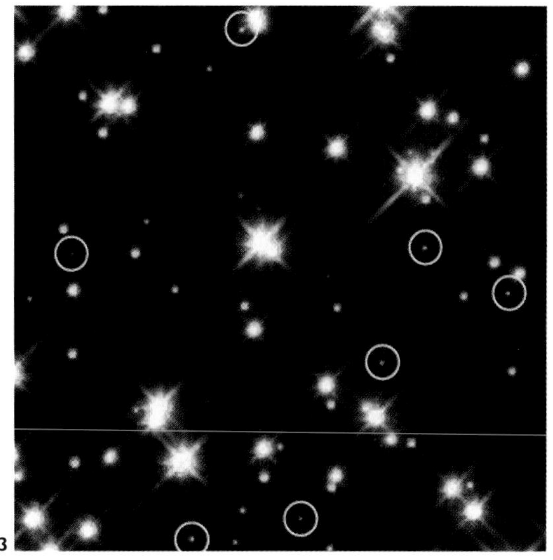

3

1. (opposite) This nebula consists of gas expelled by a hot star (near centre) which is gradually evolving into a white dwarf.

2. This image of the red supergiant star Betelgeuse reveals its huge atmosphere and a giant hotspot.

3. A region near the centre of the globular star cluster M4, with several white dwarf stars (circled above).

SEEING DOUBLE

Although most stars look like isolated points of light, closer inspection with binoculars or telescopes reveals that some are double or multiple systems that consist of two or more stars. An 'optical double' consists of two stars that appear to be very close together but which are actually at very different distances. By contrast, a 'binary' consists of two stars that revolve around each other, held together by the force of gravity. Whereas optical doubles are relatively rare, well over half of all stars are members of binary or multiple systems.

The orbital period of a binary (the time that its two stars take to revolve around each other) depends on the masses of the stars and the mean distance between them. If the period of the binary and the distance between its stars can be measured, the combined mass of the two stars can be calculated. Both stars revolve around the centre of mass of the system – a point somewhere between the two stars. If the two stars are identical, the centre of mass will be midway between them, but if one star is more massive ('heavier') than the other, it will lie closer to the more massive one. If the position of the centre of mass can be determined, and the combined mass of the two stars is known, the mass of each star can then be calculated. Observations of binaries provide the only direct way of 'weighing' the stars.

Where the two stars are far enough apart to be resolved (seen as separate points of light), the system is called a visual binary. Most binaries cannot be resolved because the stars are too close together or because one is so much fainter than the

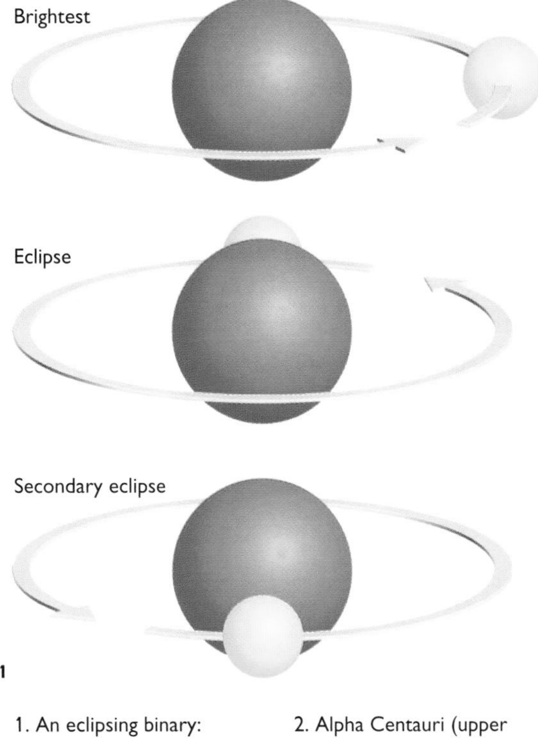

1. An eclipsing binary: two stars alternately pass in front of each other, causing variations in their combined brightness.

2. Alpha Centauri (upper left) is a binary.

3. Mizar (the brighter) and Alcor, a naked-eye binary in the Plough.

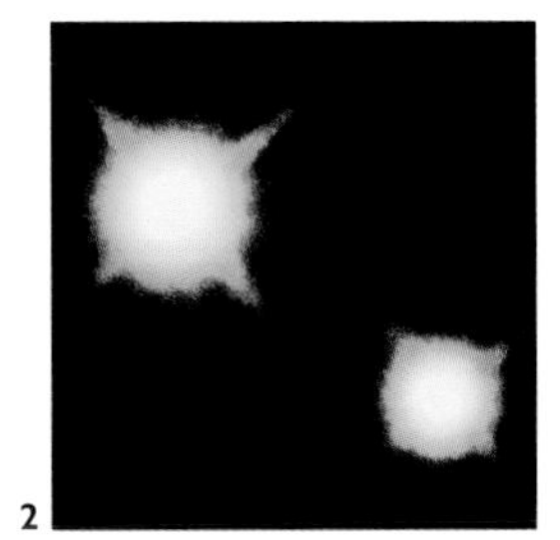

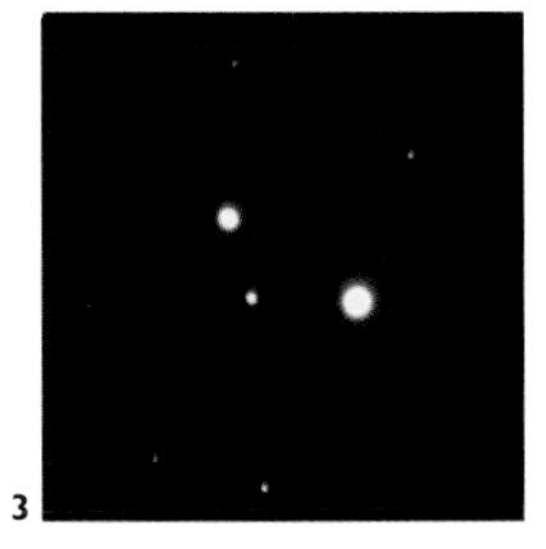

other that it cannot be seen. However, what looks like a single star may often turn out to be an astrometric, spectroscopic, or eclipsing binary.

Binaries of various kinds

In an astrometric binary, the presence of an invisible companion is revealed by its gravitational effect on the visible star. As the binary moves through space, and both of its member stars revolve around its centre of mass, the visible star will be seen to wobble from side to side. A spectroscopic binary is one in which the spectrum of what looks like a single star consists of the combined spectra of its two member stars (star A and star B). As A and B revolve around each other, each star alternately moves towards and then recedes from the Earth. Because of the Doppler effect (▷ p. 36), the lines in the spectrum of a star shift to shorter wavelengths when it is approaching and to longer wavelengths when it is receding. Consequently, the combined spectrum of the two stars contains two sets of lines that oscillate to and fro in a periodic way.

If a binary's orbit is edge-on, as seen from the Earth, each star will pass alternately in front of the other, causing a series of eclipses. What looks like a single star will dip down in brightness each time one star passes in front of the other and will return to normal after each eclipse. A binary of this kind is called an eclipsing binary. The best-known of these is Algol, in Perseus. Its main, or 'primary', eclipse, which occurs at intervals of 2.9 days, can be seen with the naked eye.

VARIABLE STARS

As their name suggests, variable stars are stars that vary in brightness. Of the many types of variable, pulsating variables are among the most common. These are stars that vary in brightness because they are expanding and contracting in a periodic way. Best known are the Cepheid variables, named after the star Delta Cephei in the constellation of Cepheus. Cepheids are yellowish giants and supergiants that brighten and fade by about 1 stellar magnitude over periods ranging from about a day up to 80 days. Other types of pulsating variable

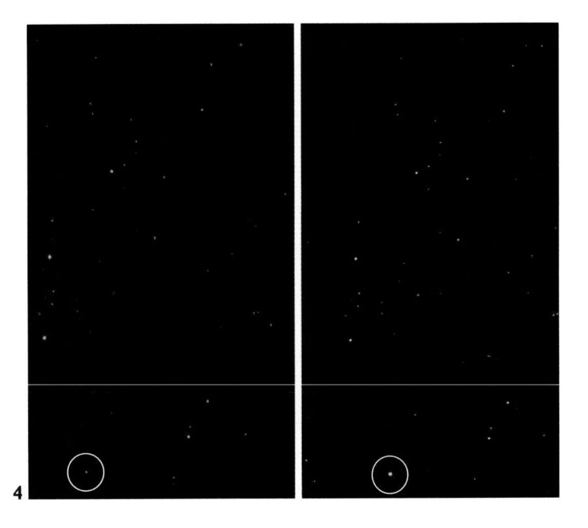

4. Two views of the long-period pulsating variable star called Mira in the constellation of Cetus (the Whale). Mira (near lower frame edges), which varies over a 332-day period is markedly brighter in the right-hand frame.

include long-period, or 'Mira', variables, and semi-regular variables. The former are cool red giants that fluctuate in brightness by factors of up to 10,000 or so over periods of 80–1000 days. A classic example is the star Mira (the Arabic word for 'wonderful'), which was so named by Arab astronomers because of its strange behaviour. At peak brightness it reaches magnitude 1.7 and is readily visible, but for much of its 330-day period it is too faint to be seen with the naked eye. Semi-regular variables have erratic periods and undergo small changes in brightness. Betelgeuse is an example of this kind of star.

Other types of variable include flare stars, novas and supernovas. Flare stars are cool red stars, which lie on the lower end of the main sequence. They display frequent outbursts, usually of a few minutes' duration, during which they brighten by a few magnitudes before dropping back to normal. These short-lived brightenings are caused by explosive eruptions, called flares, that take place on their surfaces.

A nova is a star that flares up much more dramatically, becoming, at its peak, between a thousand and a million times more brilliant than before. The rise to peak brightness may take only a few hours. Thereafter, the star's brightness

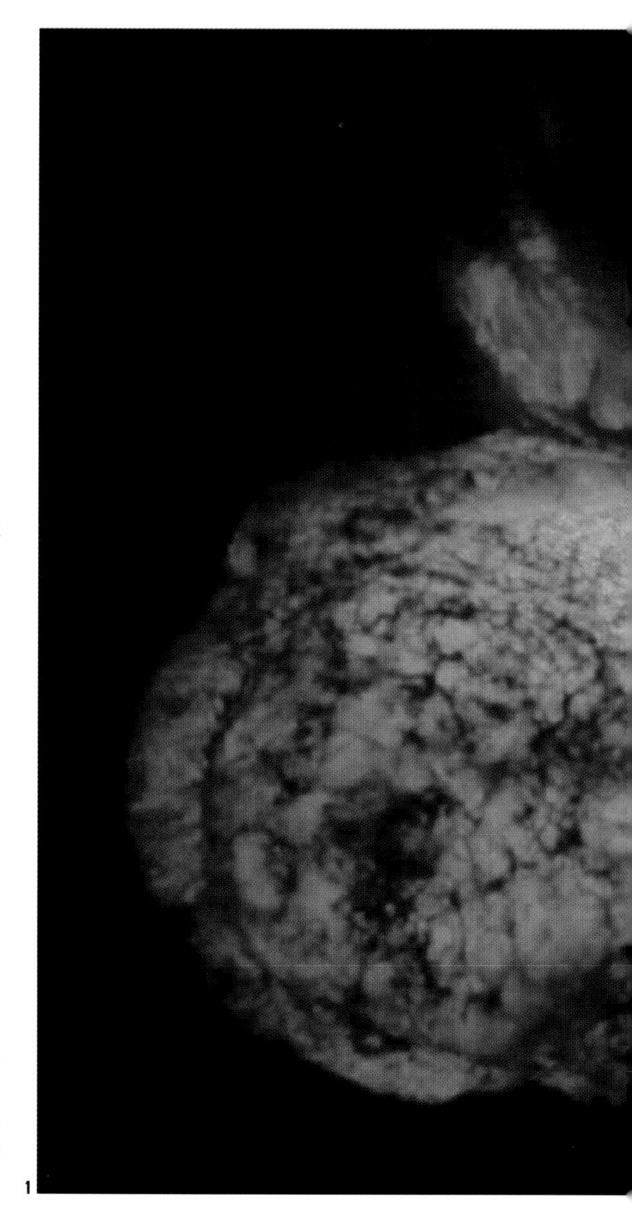

1. Eta Carinae, a supermassive star which ejected this cloud of gas and dust, is likely eventually to become a supernova.

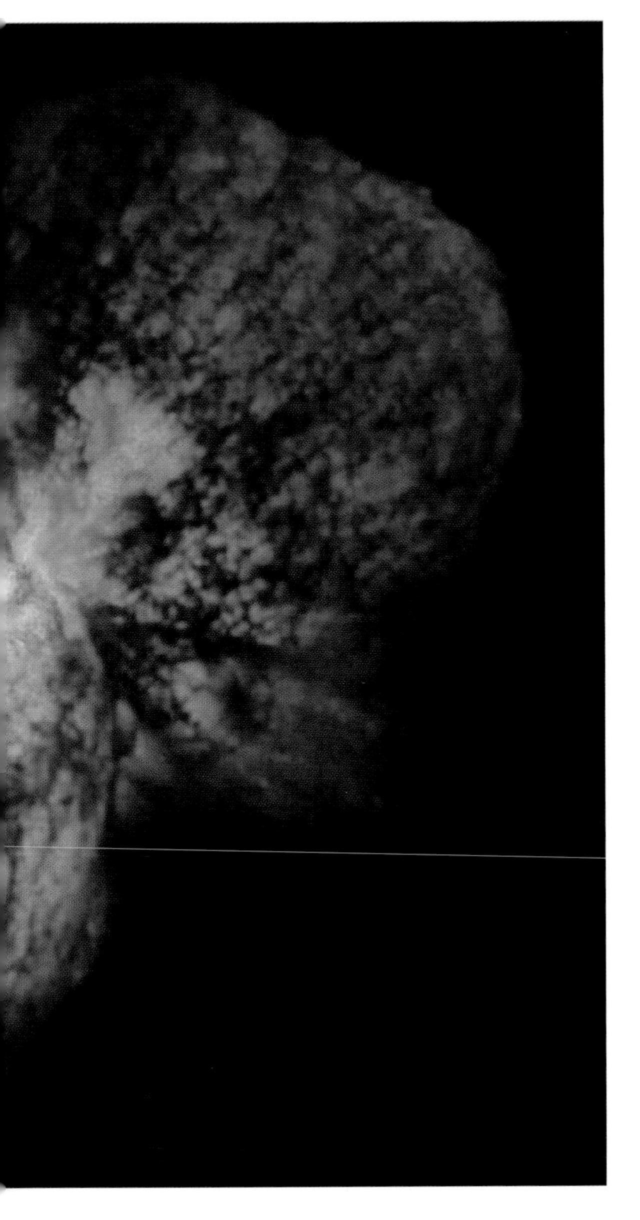

slowly fades back over months or years to its original level. The term 'nova' is rather misleading because it implies that the star is new. Before the invention of the telescope, these events seemed simply to be the appearance and subsequent disappearance of a new star because the star that flared up would have been too faint to see before its dramatic flare-up occurred. A supernova is even more dramatic. It is a true stellar catastrophe in which a star blows itself to pieces. At peak brilliancy, a supernova can, for a few days, become as bright as an entire galaxy. Astronomers' theories about the nature of these violent stellar explosions are discussed in Chapter 4.

STANDARD CANDLES

Cepheid variables are particularly important in astronomy. In 1912, an American astronomer, Henrietta Leavitt, discovered that the more luminous the Cepheid, the longer its period of variation. This period-luminosity law allows Cepheids to be used as 'standard candles' (stars of known luminosity) for measuring the distances of galaxies. Cepheids can be detected in galaxies at distances of 50–100 million light years. If their periods can be measured, their true luminosities can be found from the period-luminosity law. By comparing their apparent brightnesses with the amount of light that Cepheids with these periods actually emit, astronomers can calculate how far away they have to be in order to appear as faint as they do. This gives the distances of the galaxies within which they are embedded.

THE SUN IN CLOSE-UP

Compared to the other stars, the Sun is right on our doorstep. A huge globe of luminous gas, 109 times the Earth's diameter, and with a mass 330,000 times greater than that of the Earth, it is composed mainly of hydrogen and helium – the two lightest chemical elements. Its luminosity – the amount of energy that it pours into space every second – is 386 trillion trillion watts (3.86×10^{26} watts).

Almost all the Sun's light is radiated from a thin surface layer called the 'photosphere', which has a temperature of about 6000°C. Below the photosphere, the temperature rises rapidly to a maximum of 15 million °C at the centre of the Sun. Inside the Sun's central core, nuclei of hydrogen collide so violently that nuclear reactions weld

1

1. Taken in red light emitted by hydrogen atoms, this image of the Sun shows sunspots (dark dots), active regions (yellow areas) and filaments (red lines) – clouds of gas suspended in the solar atmosphere.

2. A composite image of the solar corona with bright active regions on the Sun's surface, dark coronal holes and solar wind.

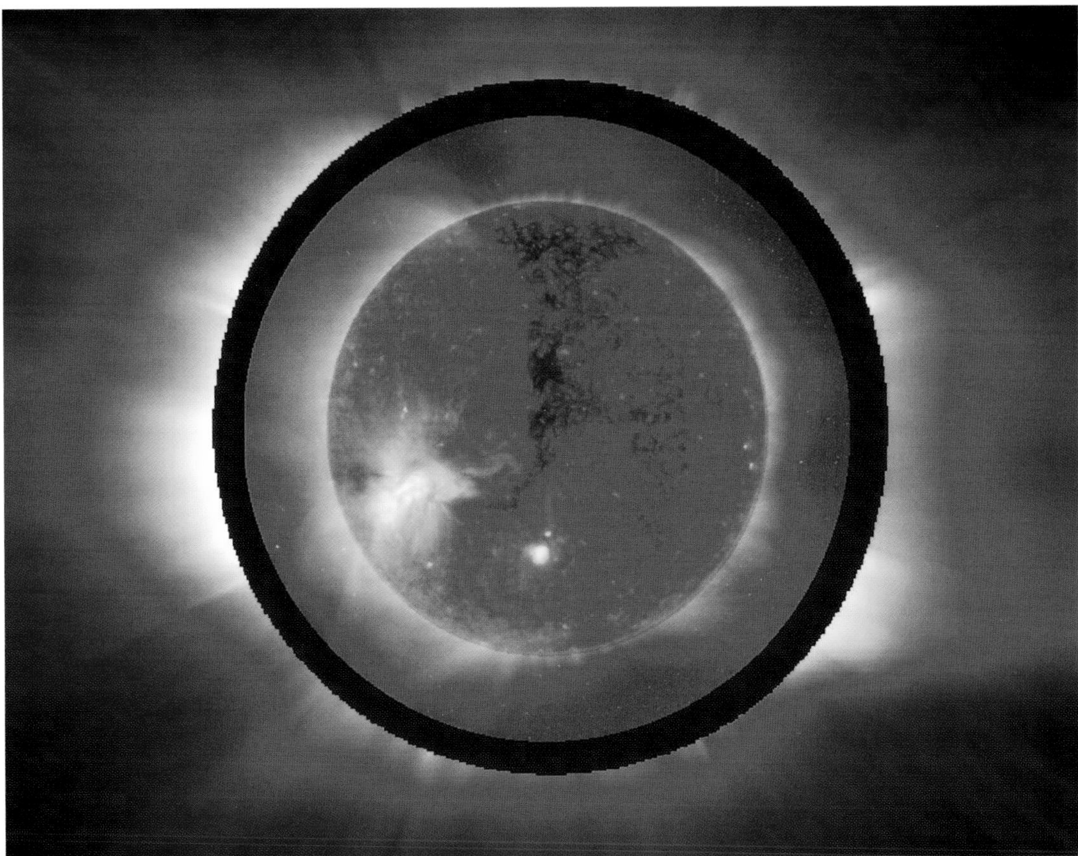

2

them together to create nuclei of the next lightest element, helium.

The mass of the resulting helium nucleus is 0.7 per cent less than the mass of the hydrogen nuclei that went into making it. According to Albert Einstein's special theory of relativity, if a quantity of matter is converted into energy, the energy released (E) is equal to the mass (m) multiplied by the square of the speed of light (c). This famous equation is usually written as $E = mc^2$. To maintain its steady output of energy, the Sun effectively destroys more than 4 million tonnes of matter every second.

Above the photosphere is a thin layer of gas called the 'chromosphere', beyond which is the corona (the Sun's outer atmosphere) which stretches out to a distance of several million

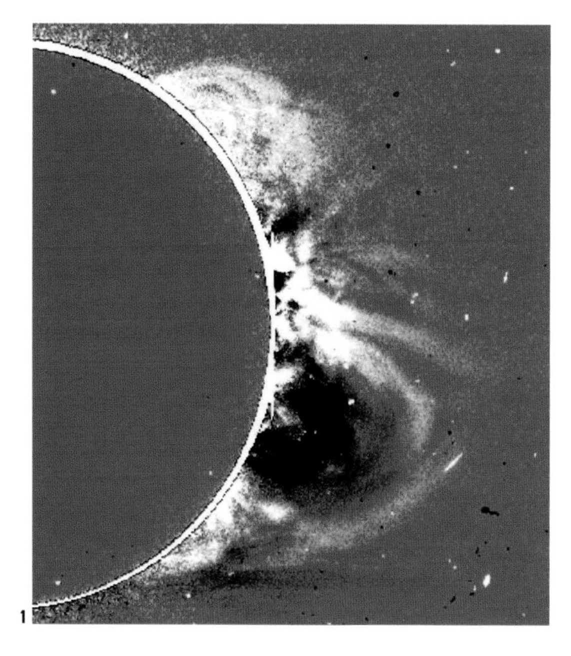
1

Between AD 1645 and 1715 – a period known as the Maunder minimum – hardly any sunspots were seen. Europe experienced a long succession of extremely cold winters during the middle part of that period.

kilometres. The temperature of the corona is between 1 and 5 million°C. Despite its very high temperature, the density of the corona is so low that, compared to the photosphere, it contains very little heat. Charged subatomic particles – mainly protons and electrons – escape from the corona into interplanetary space and flow past the Earth and planets at speeds of several hundred kilometres per second (more than a million kilometres per hour). This flow of particles is called the 'solar wind'.

Solar activity

Various forms of activity take place on the surface and in the atmosphere of the Sun. The most obvious symptom of solar activity is the presence of sunspots, patches on the photosphere which appear dark because they are cooler than their surroundings. These spots, which are often far larger than the Earth, occur when strong magnetic fields erupt through the Sun's surface. Their numbers increase and decrease in an 11-year cycle.

Other forms of activity include prominences, flares and coronal mass ejections. Prominences are huge plumes of hot gas, which surge upwards, sometimes to heights of several hundred thousand kilometres, or which hang like clouds in the solar atmosphere for weeks or months on end. Flares are explosive releases of stored magnetic energy, much of which is released in the form of X-rays and high-speed subatomic particles. A major flare can release as much energy as several billion nuclear bombs. Coronal mass ejections (CMEs) are huge bubbles of gas that surge through the corona into space.

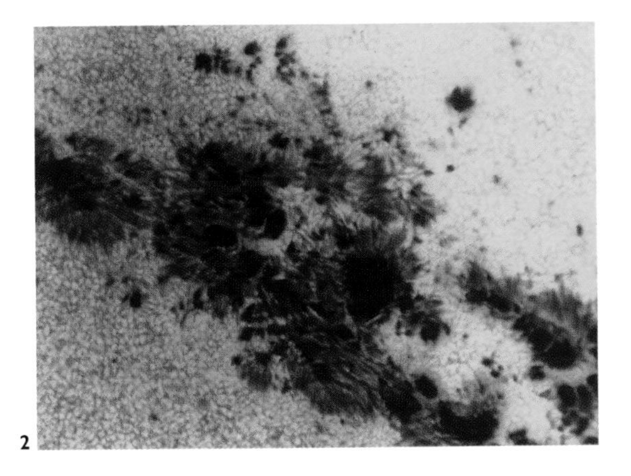

2

Flares, CMEs and disturbances in the solar wind affect the upper atmosphere and magnetic field of the Earth, disrupt radio communications, cause power blackouts and can damage electronic components on orbiting satellites.

After major solar outbursts, charged particles are catapulted into the upper atmosphere, where they cause atoms and molecules to give out light, thereby generating the Aurora Borealis and Aurora Australis – beautiful patterns of shimmering light that are often seen, respectively, in the Arctic and Antarctic skies.

1. This image from the SOHO spacecraft, shows huge eruptive prominences – plumes of hot gas rising out of the Sun's atmosphere (the corona).

2. Each major spot in this sunspot group has a dark core (umbra) and a less dark outer region (penumbra).

3. Ultraviolet image of an eruptive prominence 560,000 km (348,000 miles) high. False colours show the different brightnesses.

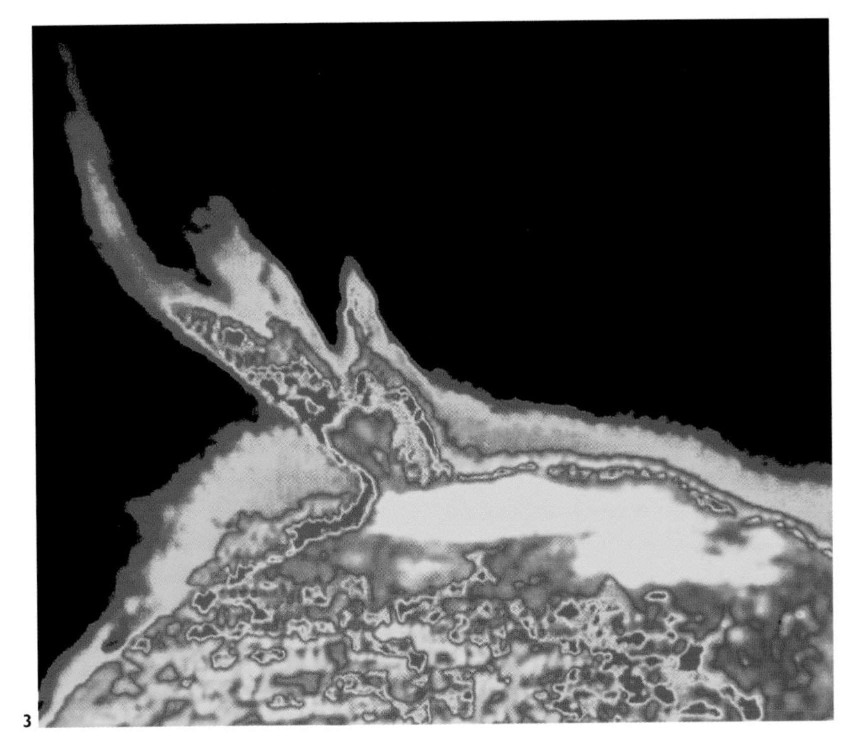

3

3

LIFE CYCLES OF STARS

3 LIFE CYCLES OF STARS

Stars do not live for ever. They are born inside clouds of gas and dust, develop to maturity, then grow old and 'die'. Powered throughout their lives by nuclear reactions that take place deep down in their interiors, they eventually run out of fuel. At the end of their lives, most stars simply fade away, but some blow themselves to pieces in a final act of self-destruction. Because the main stages in the evolution of a star last for millions or billions of years, we cannot watch an individual star pass through its various stages of life. However, when we look at a large sample of stars, we see some that are very young, some that are mature and others that are old.

By studying large numbers of stars at different stages in their lives, astronomers can discover the various stages through which an individual star will pass in the course of its lifetime. In this way they have been able to piece together the life histories of stars in general.

Previous page: The Orion Nebula is a glowing cloud of gas lit up by several hot young stars embedded within it. Located at a distance of 1500 light years, it is part of the nearest star-forming region to the Sun.

RAW MATERIALS

Tenuous clouds of gas and dust, which permeate interstellar space (the space between the stars), provide the raw materials from which stars are made. Some of these clouds emit visible light. Known as emission nebulas (the name 'nebula' derives from the Latin word for 'mist'), these clouds, which are composed mainly of hydrogen, shine because they contain one or more highly luminous stars of spectral type O or B0 (▷ p. 36). These stars are so hot that they emit large amounts of energetic ultraviolet light, which ionizes (knocks electrons out of) some of the hydrogen atoms in those clouds. When electrons are recaptured, they release energy in the form of visible light. As an electron drops down to lower-energy orbits closer to the atomic nucleus around which it is revolving, it emits light of particular wavelengths. The spectrum of an emission nebula consists of several bright lines, each with its own wavelength.

The best-known example of an emission nebula is the Great Nebula in Orion. Also known as M42 (object number 42 in the catalogue of 'fuzzy ▷▷

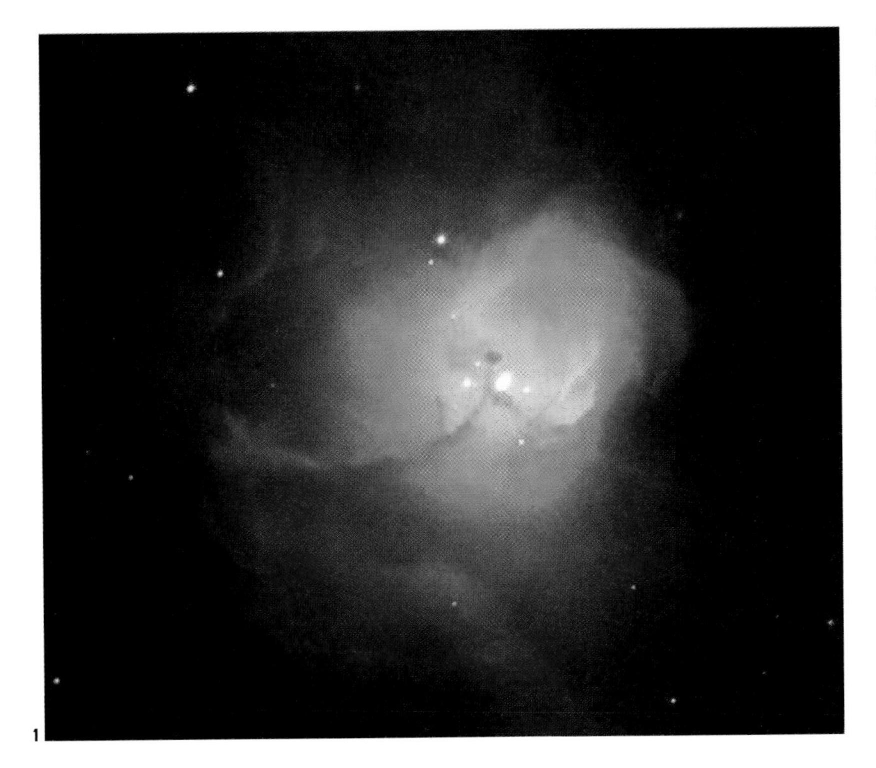

1. This batch of hot young stars, embedded in their embryonic cloud of glowing gases, is located in the Small Magellanic Cloud, a galaxy which is about 190,000 light years from Earth. Each of these stars is about 300,000 times as brilliant as the Sun.

Previous page: Silhouetted against a red emission nebula is the Horsehead Nebula, part of a cloud of gas south of Orion's belt.

1. The blue nebulosity surrounding this group of stars, which lies north of the Orion Nebula, is starlight scattered (reflected) by dust.

objects' published in 1781 by the French astronomer Charles Messier), it is located in the 'sword' of Orion, a group of faint stars that hangs below (south of) Orion's belt. Under good conditions, the nebula can be glimpsed with the naked eye. A small telescope will show the compact group of four hot young stars, called the Trapezium, which is responsible for causing the nebula to glow. The Orion Nebula lies at a distance of 1500 light years, is nearly 20 light years in diameter and is the nearest object of its kind.

Cool gas and dusty grains

If a gas cloud does not contain a brilliant hot star, it will not shine. However, when light from a background star passes through an interstellar cloud, atoms within the cloud absorb light at particular wavelengths and therefore superimpose additional dark lines on the spectrum of the star. Lines of this kind, which are called interstellar lines, were first identified in the spectrum of the star Delta Orionis (Mintaka), a spectroscopic binary. Whereas the two sets of lines in the spectrum of this binary shifted to and fro in wavelength in a periodic way, there were other lines that remained fixed in wavelength. Astronomers realized that

these lines must have been added to the spectrum of the binary as its light passed though a gas cloud.

Interstellar dust makes its presence felt in a number of ways. If a cloud contains enough dusty particles to absorb all or most of the light heading our way from stars that lie behind it, it will show up as a dark patch against the stars. A cloud of this kind is called a dark (or absorption) nebula. Classic examples include the Horsehead Nebula, in Orion, and the Coalsack, a near-circular dark cloud that is visible to the naked eye in the Southern Cross.

The dust grains are tiny, only about 100–1000 nanometres in diameter. Because they are similar in size to the wavelengths of visible light, they are very effective at scattering (reflecting) and absorbing starlight (longer wavelengths, such as infrared and radio, are much less affected). Consequently, interstellar dust causes distant stars to appear fainter than they should. Since short-wave (blue)

Gas in the Orion Nebula is so thinly distributed that a cylinder, 1 m (3 ft) in diameter and extending all the way through the nebula (nearly 20 light years) would contain less material than a 1-kg (2-lb) bag of sugar.

light is affected more than long-wave (red) light, dust also causes distant stars to look redder (or less blue) than they really are. Furthermore, where one or more bright stars lie in front of, or adjacent to, clouds of dust, the dust grains scatter some of their light towards the Earth. This causes those stars to be surrounded by hazy patches of bluish light, which are called reflection nebulas.

Cool clouds composed of ordinary hydrogen atoms (neutral hydrogen) emit radio waves with a wavelength of 21.1 cm (8 in). Molecules (a molecule consists of two or more atoms joined together) also emit and absorb radiation, mainly at infrared and microwave wavelengths. Over the past few decades, astronomers have discovered about 100 species of molecule in space. Most are organic molecules, which consist of the element carbon combined with, predominantly, hydrogen, nitrogen, or oxygen. Examples include molecular hydrogen (H_2), formaldehyde, ethyl alcohol and at least one amino acid. The most basic building blocks of life exist in large quantities in interstellar space.

Clouds of hydrogen and helium that are rich in molecules are known as molecular clouds. Typically up to 100 light years in diameter, clouds of this kind have masses of up to a hundred thousand times the mass of the Sun. Giant molecular clouds (GMCs) are even larger, with diameters of up to 300 light years and masses ranging from a hundred thousand to several million solar masses. As starlight – especially energetic ultraviolet light – breaks the fragile bonds that hold molecules together, the best conditions for the formation and survival of complex molecules occur in massive dust-laden clouds where the density is relatively high (this makes it easier for molecules to form), temperatures are low (a few degrees or tens of degrees above absolute zero), and the dust particles shield the molecules from the destructive effects of starlight.

HYDROGEN ANNOUNCES ITS PRESENCE

Clouds of hydrogen atoms emit radio waves, with a wavelength of 21.1 cm (8 in), which penetrate our atmosphere and are detectable by ground-based radio telescopes. In a hydrogen atom, when the orbiting electron and the central proton spin in the same direction, the atom has slightly more energy than when they spin in opposite directions. When the electron flips from spinning in the same direction to spinning in the opposite direction, the change in energy is emitted as radiation with a wavelength of 21.1 cm and a frequency of 1.42 gigahertz. The existence of this radiation, predicted in 1944 by Dutch astronomer H. C. van de Hulst (right), was detected in 1951. By studying 21.1-cm emissions, astronomers can map hydrogen clouds in our own galaxy and in others.

HOW STARS ARE BORN

A star begins to form when a cloud of gas and dust starts to collapse under its own weight, a process that can get started only if the inward-acting pull of gravity is sufficiently powerful to overcome the cloud's internal pressures. Since the pressure exerted by a gas depends upon its temperature, the hotter the cloud, the less likely it is to collapse. However, if a cool cloud is sufficiently massive, or if a sufficiently dense clump, or 'core', forms inside a large cloud, gravity will win the battle and the cloud or core will begin to collapse.

Although most of the interstellar gas is too diffuse for this to happen, there are various ways in which a cloud can be squeezed sufficiently for star birth to begin. For example, a cloud will be compressed when it enters one of the spiral arms of our galaxy (where the concentration of matter is greater) or if it collides with another cloud.

1. In this false-colour infrared image of part of the Omega Nebula (Messier 17), a star-forming region in the constellation of Sagittarius, young stars that are heavily obscured by dust are shown in red.

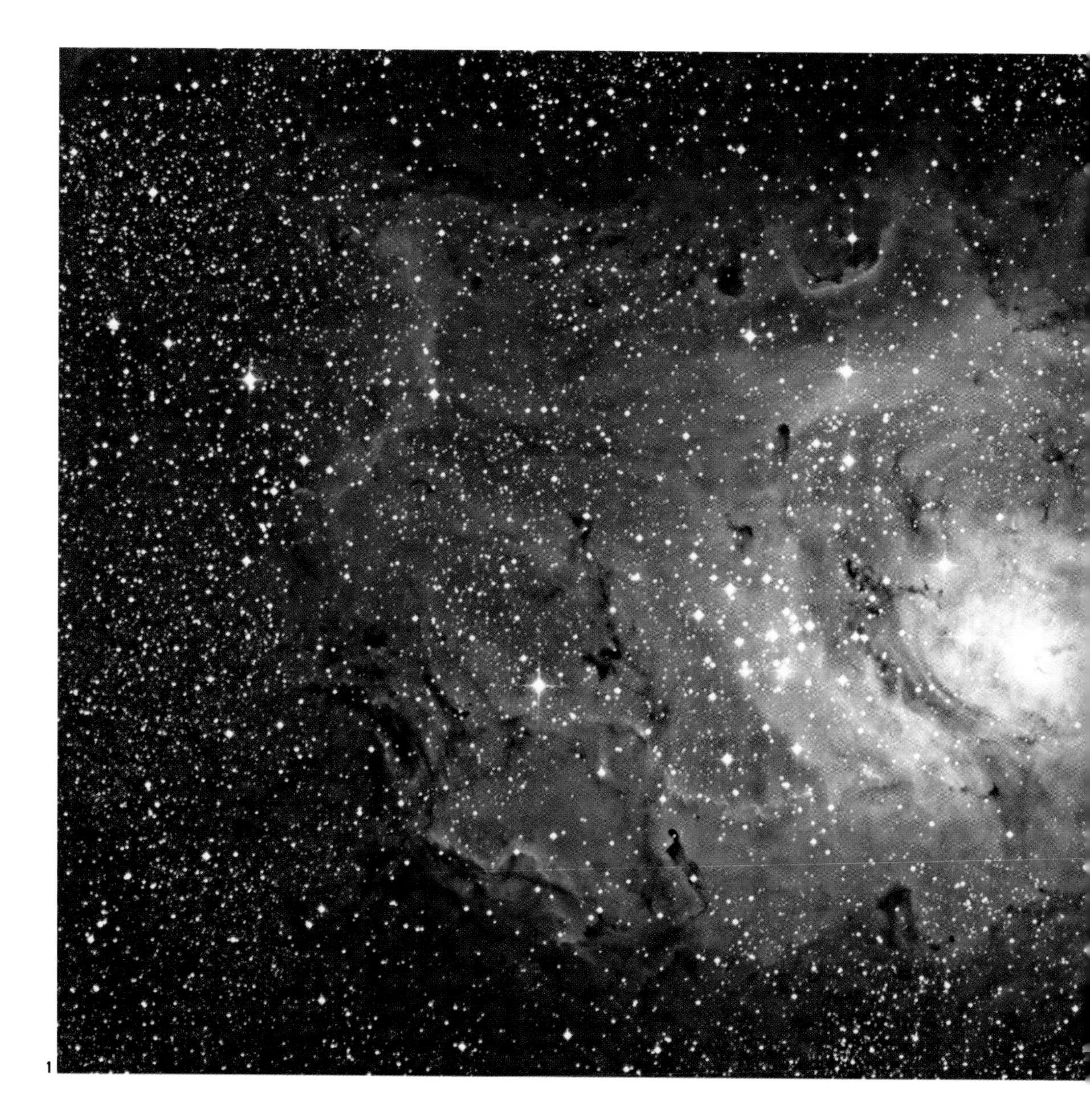
1

Likewise, dense clumps of gas can be formed when the shock wave from a supernova explosion, or the strong stellar wind (a fast-moving flow of gas) from a highly luminous star, ploughs into an interstellar cloud. The best conditions for star formation occur in giant molecular clouds, where temperatures are low and densities are relatively high.

When it starts to contract, an individual cloud core is typically about a light year or so in diameter and is rotating very slowly. As it falls in on itself, it spins round faster and faster, just as twirling ice-skaters spin slowly when their arms are outstretched, but much faster when they draw their arms in tightly to their bodies. If it ends up rotating too rapidly, it breaks into two or more fragments, which will continue to revolve round each other and will give birth to a binary or multiple system of stars. Otherwise, the inner part of the nebula collapses into a concentrated blob of gas (a protostar) and the rotational motion flattens out the rest of the infalling gas and dust to form a disc or lens-shaped nebula around the protostar. As the protostar continues to contract and sucks in more matter from the surrounding nebula, its density and temperature rise. Collisions between atoms and molecules become more frequent and violent,

1. The Lagoon Nebula in Sagittarius contains a recently formed cluster of stars. Star formation continues in the brightest part of the nebula.

pressure builds up and the rate of contraction slows down. Visible and ultraviolet radiation from the hot interior of the protostar is absorbed by grains of dust, and heats them to temperatures of a few hundred degrees. Although the protostar itself is hidden from view by the dusty cocoon that surrounds it, infrared radiation emitted by the heated dust reveals its presence.

The end of the beginning

When the temperature at the centre of the forming star reaches about 10 million °C, nuclear fusion reactions commence in its core, converting hydrogen into helium and releasing copious amounts of energy. Pressure exerted by the hot gas inside the new-born star halts its contraction, and it reaches a state of balance, where the pressure inside is sufficient to prevent gravity compressing it any further. When it reaches this stage in its development, it has become a main sequence star.

A strong stellar wind (an outflow of gas from the new-born star) soon blows away the remaining gas and the shroud of dust within which the young stellar object was hidden, so enabling the new-born star to be seen directly.

High-mass protostars collapse and heat up faster than low-mass ones. Collapsing clouds that are 10 times as massive as the Sun take less than 100,000 years to turn into main sequence stars. Stars with masses similar to the Sun take several million years to reach this stage, and stars at the lower end of the main sequences take tens of millions of years. The more massive the star, the hotter and more luminous it will be. Newly formed stars with more than 10 times the mass of the Sun join the top end of the main sequence, whereas stars with one-tenth of the Sun's mass arrive near the bottom end.

Protostars with less than 8 per cent of the mass of the Sun (less than 80 times the mass of Jupiter) never achieve high enough temperatures for sustained hydrogen fusion reactions to switch on. They fail to become genuine stars. Instead, they become brown dwarfs – dense gaseous bodies that shine dimly because they are radiating away stored heat accumulated while they were contracting. A newly formed brown dwarf is likely to have about

☆ The coolest known brown dwarf, Gliese 570D, has a surface temperature of about 480°C (896°F), comparable to the temperature on the surface of the planet Venus.

1. (opposite) A jet of ejected material (lower centre) is heading downwards from the protostar HH-34.

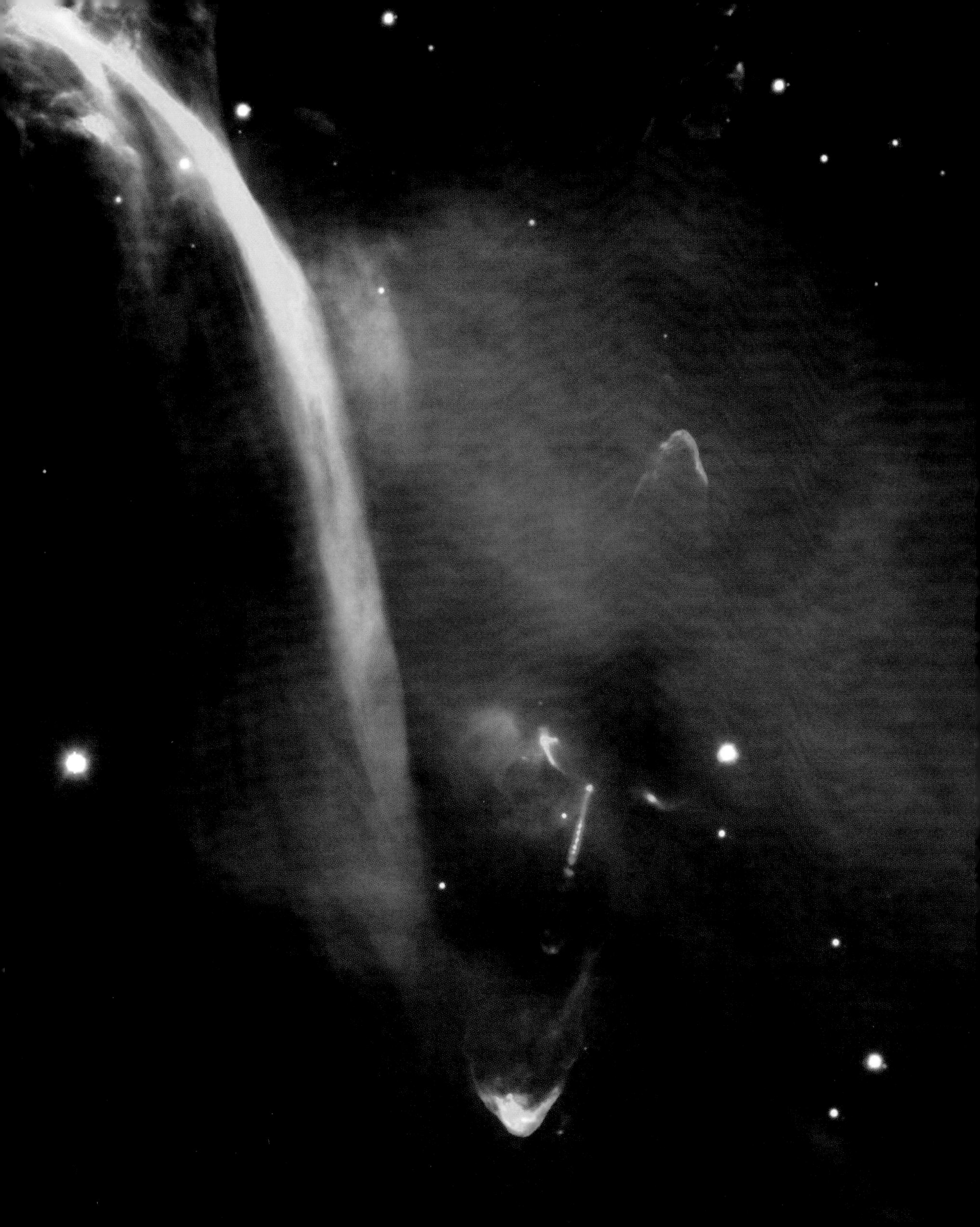

one-ten-thousandth of a solar luminosity and a surface temperature of 2000–2500°C. As it grows older, it will cool down and fade. Because of their low luminosities, brown dwarfs are hard to detect, and it is only in the past few years that astronomers have begun to find them in large numbers.

New stars tend to form in batches, where a number of dense cores form within a giant molecular cloud. The nearest stellar nursery of this kind is the Orion molecular cloud, a collection of dust-laden clouds of which the visible Orion Nebula is a small part. Infrared observations show that a new batch of stars is forming at present in a 10,000-solar mass cloud called OMC-1, which lies behind the visible nebula. Images obtained by the Hubble Space Telescope show that the Orion region contains hundreds of young and newly forming stars, many of which are surrounded by flattened or doughnut-shaped clouds of gas and dust.

Many astronomers believe that dust-laden clouds like these, which are known as proto-planetary discs (or 'proplyds'), contain the material from which planetary systems are formed, and that planet formation is a natural by-product of the way in which stars themselves are born. While no planet around another star has yet been observed directly, since 1995 astronomers have been able to detect planets with masses comparable to that of the planet Jupiter by measuring the microscopic wobbles that their gravitational attractions impart to their parent stars. By the autumn of the year 2000 about 50 planets had been discovered revolving around main sequence stars similar to the Sun.

HOW PLANETS ARE FORMED

Many astronomers believe that in the flattened nebula that surrounds a new-born star, grains of dust fall together in clumps, or stick together one by one, until they have grown into bodies called planetesimals, which are 5–10 km (3–6 miles) in diameter. Collisions between planetesimals cause some to break into fragments and others to grow. Larger ones mop up the smaller ones, eventually forming bodies the size of the Moon or the planet Mercury. Subsequent collisions build up Earth-like planets and the rocky cores of giant planets. In the cooler outer parts of the nebula, these rocky cores sweep up large quantities of gas to form bodies such as Jupiter (right) and Saturn. Dusty discs have been observed around a number of young main sequence stars. A few of these discs contain ring-like gaps, regions that may have been swept clear of dust by a recently formed planet or by a planet that is in the process of being formed.

LIFE ON THE MAIN SEQUENCE AND BEYOND

Once it has joined the main sequence, a newly formed star has attained a state of balance whereby the hot gas in its interior exerts sufficient pressure to counterbalance the gravitational force that is trying to make it contract. This state of balance will be maintained for as long as the star is capable of generating energy at a steady rate and is radiating energy away from its surface at the same rate as it is being produced in its core. If the rate of energy generation were to increase, the star would expand until it was again in a state of balance. Conversely, if it were to decrease, the star would shrink until its internal pressure could once again halt its gravitational contraction.

While on the main sequence, a star is powered by nuclear reactions that convert hydrogen to helium and release energy – reactions similar to those that take place inside the Sun. Astronomers often refer to this process as 'hydrogen burning', a convenient, though somewhat misleading, term (nuclear fusion has nothing to do with 'burning' in the normal sense of the word). As hydrogen burning continues, more and more of the stock of

1. The Sun is a typical main sequence star. Dust suspended in the Earth's atmosphere causes the setting Sun to appear orange or red in colour.

2. This globular cluster of stars, called Hodge 11, is associated with the Large Magellanic Cloud, a nearby galaxy.

1. Omega Centauri is a globular cluster which contains about a million stars. Located at a distance of 17,000 light years from Earth, it is one of the oldest objects in our galaxy.

2. Relative sizes of different types of star. (a) The Sun (right) compared to a red giant (left); (b) a white dwarf compared to the Sun; (c) a neutron star (right) compared to a white dwarf (left); (d) a black hole with a mass similar to the Sun (below) compared to a neutron star (above).

hydrogen in the star's core is converted into helium, and its reserves of hydrogen 'fuel' diminish. When practically all the available hydrogen has been consumed, hydrogen burning ceases, and the core can no longer support the weight of the rest of the star. Major changes then occur, which cause the star to leave the main sequence.

The time that a star spends in its main sequence stage (its main sequence lifetime) depends on its mass. Strangely enough, the higher its mass, the shorter its lifetime. The explanation for this apparently peculiar state of affairs lies in the link between a star's mass and its luminosity. As a rough rule of thumb, the luminosity of a main sequence star (expressed in solar luminosities) is approximately equal to the cube of its mass. For example, if a star is twice as massive as the Sun, it will be eight times as luminous ($2^3 = 8$), or if it is 10 times as massive, 1000 times (10^3) as luminous (in fact, a 10-solar mass star is several thousand times more luminous than the Sun). Although a more massive star has more fuel, it has to burn that fuel at a rapid rate to sustain its very high luminosity.

The Sun is believed to have enough fuel to keep it shining as a main sequence star for a total of 10 billion years. It is just under 5 billion years old at present, and is believed to have enough hydrogen fuel to keep it shining for at least another 5 billion years. A star with 10 times the Sun's mass will have 10 times as much fuel but will be consuming it several thousand times faster, so will use it up in a few tens of millions of years. The most massive stars have main sequence lifetimes of no more than a few million years. By contrast, a star with one-tenth of the Sun's mass and one-thousandth of its luminosity will outlive the Sun 100 times over.

The first white dwarf to be identified was Sirius B, a faint companion to the brightest star. In 1914, American astronomer Walter S. Adams showed that it was comparable in size to a planet.

Approaching old age

When a star's core runs out of hydrogen, and ceases to generate energy through hydrogen fusion reactions, it begins to contract, squeezed inwards by the great weight of the star's outer layers. As it contracts, its temperature increases, and the temperature in a thin shell of gas around the core rises high enough for hydrogen-burning reactions to take place there. As the core contracts further, the temperature in the shell rises further, the reactions proceed at a faster rate, and the output of energy from the shell rises so greatly that the luminosity of the star increases well beyond its main sequence value. This upsets the balance inside the star and causes it to expand. As the radius increases, the surface area increases rapidly (if the radius is doubled, the surface area is quadrupled) so that, although the luminosity increases, the surface temperature decreases. Plotted on an H-R diagram (▷ p. 37), the star moves up and to the right, away from the main sequence and towards the red giant region.

While the star is evolving away from the main sequence, helium produced in the shell is added to

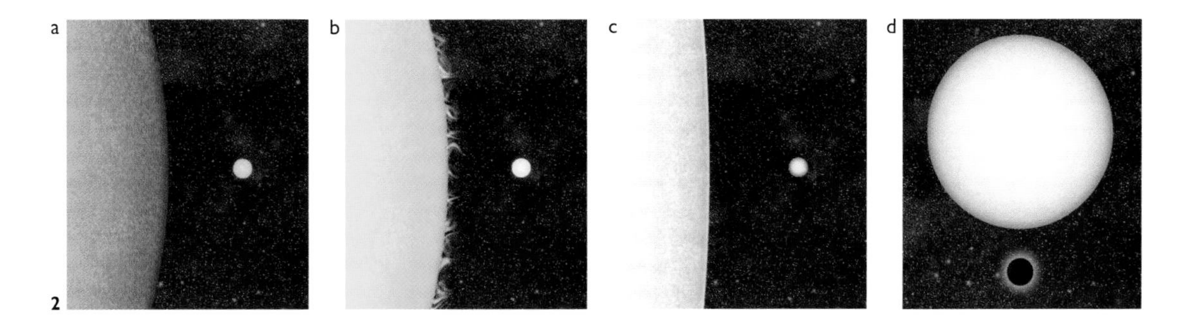

1
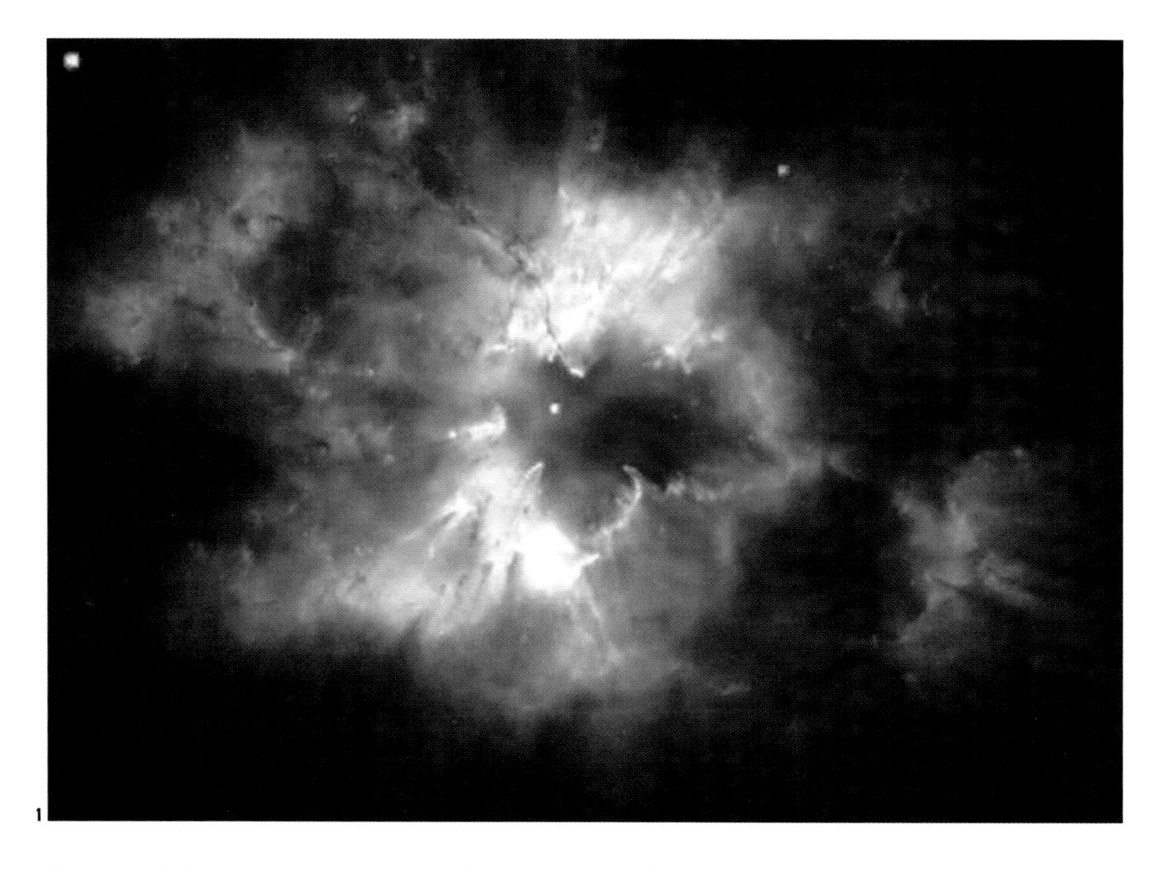

the core, which continues to contract and become progressively hotter. Eventually, when the core temperature reaches about 100 million degrees, it is hot enough for a helium-burning reaction to begin. This process converts helium into carbon and oxygen, with the release of energy. The reaction is called the triple alpha reaction because it involves the welding together of three helium nuclei (which, traditionally, were called alpha particles) to form two nuclei of carbon. A carbon nucleus can then combine with another helium nucleus to form a nucleus of oxygen.

When helium burning begins in its core, the star settles down to another stable phase of existence as a red giant. But because it is so luminous, a red giant consumes its reserves of helium fuel rapidly. For a star comparable to the Sun, the red giant phase will last for a few hundred million years at most. A high-mass star remains in its red giant (or red supergiant) stage for a much shorter time.

DEATH OF A SUN-LIKE STAR

When a star comparable to, or less massive than, the Sun has used up all the helium fuel in its core, it cannot generate any more energy. Towards the end of its red giant phase it puffs off a shell, or a series of shells, of unburned hydrogen into space to form a glowing cloud of gas called a planetary nebula. The name is misleading – planetary nebulas have nothing to do with planets – but was given by the 18th-century astronomer William Herschel, who thought these hazy blobs of light looked like the discs of planets. An example is the Ring Nebula (M57) in the constellation of Lyra. It looks like a ring around its faint central star, but is actually an elongated cylinder of gas seen end-on. Many planetary nebulas consist of two or more concentric shells of gas, and some have more complex structures.

As the cool envelope of gas drifts away from the dying star (a solar-mass star may lose up to 40 per

1. The planetary nebula NGC 2440 is a shell of gas that surrounds one of the hottest-known white dwarfs. The white dwarf is the bright dot near the centre.

2. The Ring Nebula (M57), a planetary nebula in the constellation of Lyra, is a barrel-shaped cloud of gas. Since, from Earth, we are looking at the cloud end on, light emitted by the 'walls' of the 'barrel' creates a ring-like impression.

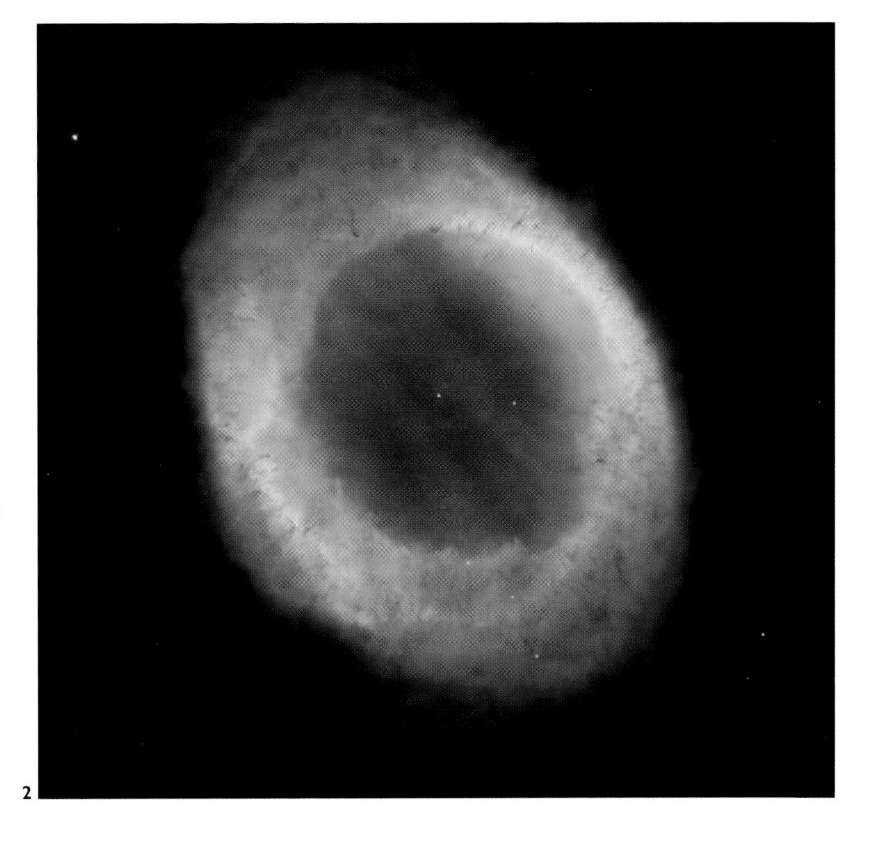

cent of its mass at this stage), it exposes the hydrogen- and helium-burning shells, which soon cease to generate energy. With no energy sources to support it, the star shrinks rapidly, its surface temperature rising to over 30,000°C, sometimes to above 100,000°C. Ultraviolet radiation from the hot star surface ionizes the expanding shell of gas and causes it to emit visible light, so giving rise to the visible planetary nebula. However, after shining for only a few tens of thousands of years, this will eventually fade and merge with the interstellar gas.

Because the shrinking core of a solar-mass star cannot contract enough to raise its temperature for carbon-burning fusion reactions to begin, no further nuclear energy generation is possible. The remnant of the dying star stops contracting when the pressure exerted by fast-moving electrons becomes high

1

SUN AND EARTH – THE FINAL CURTAIN

So far as the fate of the Sun and planet Earth is concerned, the long-term future is bleak. By the time the Sun has become a red giant – in about 5 or 6 billion years' time – it will have expanded to between 50 and 100 times its present diameter, and its luminosity will have increased a thousandfold. Mercury, and possibly Venus, will have been swallowed up by the distended Sun, and the temperature at the surface of the Earth will have risen to more than 1000°C. The oceans will evaporate, the atmosphere will be driven off into space, the surface rocks will melt, and life as we know it will become impossible on planet Earth. At the end of its red giant phase, the Sun will shed its outer shell of gas and will shrink down – within a few tens of thousands of years – to become a white dwarf 1000 times fainter than the present-day Sun. By the time the Sun has become a white dwarf, the Earth, or what is left of it, will have cooled to a frozen relic, and life will no longer be sustainable on any of the worlds in the Solar System.

enough to resist the crushing force of gravity. By now, the star is comparable in size to the Earth, has a luminosity of about one-thousandth that of the Sun, and is composed mainly of carbon and oxygen nuclei packed so closely that the mean density of the star's material is hundreds of thousands of times denser than water. The star has become a white dwarf.

With no nuclear energy sources to sustain it, a white dwarf gradually radiates away its immense reserves of heat. The pressure exerted by its electrons does not change as its temperature drops, so the dying star cannot shrink any further. It slowly cools and fades, and after billions of years will become a cold, dark body called a black dwarf. (No star in our Galaxy has yet reached that stage.)

Stars of medium mass (five times the mass of the Sun) evolve off the main sequence much sooner (after 100 million years or so). After reaching the red giant phase, they become unstable and start to pulsate, becoming, for a time, a pulsating variable, such as a Cepheid or long-period variable (▷ p. 43). When the helium core has been converted to carbon and oxygen, it contracts. The temperature in the shell of gas rises, so helium-burning reactions can occur there, and the star swells again, possibly becoming a supergiant, before expelling its outer layers, creating a planetary nebula and evolving into a white dwarf. For a high-mass star, the story is much more dramatic and is told in Chapter 4.

1. The Dumbbell Nebula (M27) in the constellation of Vulpecula (the Fox) is a planetary nebula with a distinctive hour-glass shape.

2. Located at a distance of 450 light years, the Helix Nebula, in the constellation of Aquarius, is the nearest planetary nebula to Earth. Its apparent diameter is about half the apparent size of the Moon in the sky.

2

4

EXPLODING STARS AND REMNANTS

4 EXPLODING STARS AND REMNANTS

In 1931, the Indian astrophysicist Subrahmanyan Chandrasekhar showed that the maximum possible mass that a white dwarf could possess was 1.4 times the mass of the Sun. If a dying star, which has consumed all its available nuclear fuel, has a mass that exceeds this value – which is known as the Chandrasekhar limit – the crushing force of its own gravity will overwhelm even the immense pressure exerted by fast-moving electrons. Towards the ends of their lives, most medium-mass stars lose enough material into space to bring their final masses below this 'weight limit', and they end their days in unspectacular fashion as slowly fading white dwarfs. For stars which have masses in the region of 10–100 solar masses the end is much more dramatic. Their cores collapse to form incredibly dense neutron stars, or enigmatic black holes, while the rest of their material is blasted into space in catastrophic supernova explosions.

Previous page: Part of the Vela supernova remnant, the remains of a star that exploded some 12,000 years ago. Wisps of nebulosity reveal where the expanding cloud of debris is colliding with interstellar gas.

END STAGES FOR HIGH-MASS STARS

In high-mass supergiant stars, the temperature of the shrinking core continues to rise, and different nuclear fuels ignite in turn. When the core temperature reaches about 600 million°C, carbon-burning reactions produce elements such as neon, magnesium and oxygen. When its reserves of carbon have been exhausted, the core shrinks further, its temperature rises higher still, and further reactions take place, leading eventually to the production of elements such as sulphur, silicon and, finally, iron. Each successive reaction sustains the star for a shorter and shorter time. For a star of 25 solar masses it has been calculated that helium burning will last for 500,000 years, carbon burning for 600 years, oxygen-burning reactions for about six months and silicon burning for about one day.

To support itself, a star relies on reactions that give out energy. But to weld iron nuclei together, energy has to be put in. Once the core has been converted into iron, it cannot support itself, so it collapses virtually instantaneously. If the mass of the collapsing iron core exceeds the Chandrasekhar limit (1.4 solar masses), it cannot become a white dwarf. Gravity will overwhelm the pressure exerted by its fast-moving electrons. As the matter in the collapsing core is squeezed to incredible densities, positively charged protons combine with negatively charged electrons to form electrically neutral neutrons. In the process, vast numbers of neutrinos are released – ghostly particles with zero electrical charge and zero, or exceedingly tiny, masses. ▷▷

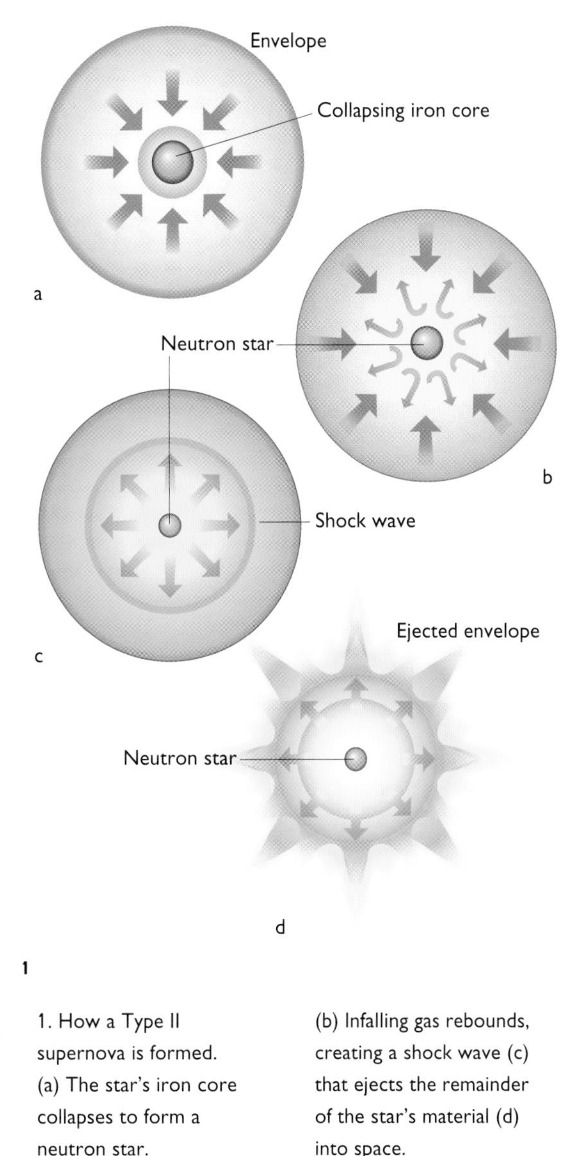

1. How a Type II supernova is formed. (a) The star's iron core collapses to form a neutron star. (b) Infalling gas rebounds, creating a shock wave (c) that ejects the remainder of the star's material (d) into space.

Previous page:
The Crab Nebula is the expanding remnant of a star that was seen to explode by Chinese astronomers in the year 1054.

1. This image shows X-ray emissions from the supernova remnant Cassiopeia A. The bright spot near the centre may be a neutron star.

Provided that the mass of the collapsing core does not exceed 2–3 solar masses, the pressure exerted by close-packed neutrons will prevent it collapsing any further once its density has risen to about 400 million million times greater than the density of water (about 4×10^{17} kg per cubic metre). By this stage – a few tenths of a second after it began to collapse – the core has shrunk to a radius of about 10 km (6 miles) and has become a neutron star, a body so compressed that a teaspoonful of its material, if brought to Earth, would weigh nearly a billion tonnes.

A star explodes

When infalling material from the main body of the star collides with the rigid new-born neutron star at its centre, it rebounds and drives out a powerful shock wave (a gigantic version of the wave produced by a supersonic aircraft), which blasts most of the star's material into space. The expanding shock wave reaches the surface of the doomed supergiant star a few hours after the collapse of its core and propels its photosphere (visible surface) outwards at a speed of around

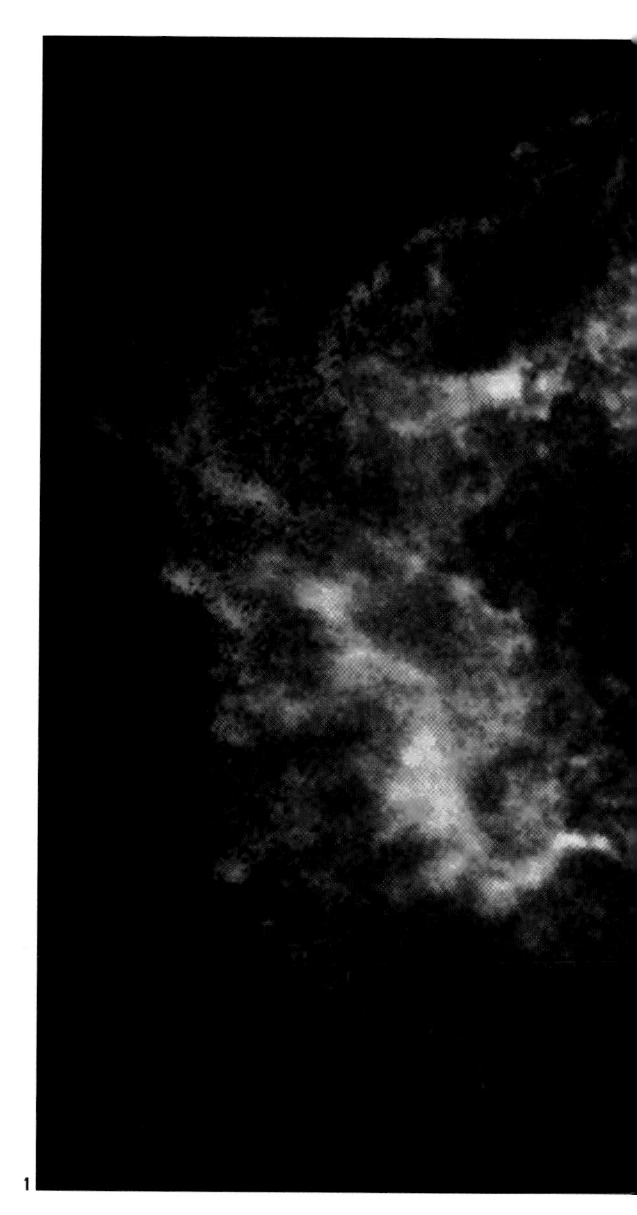
1

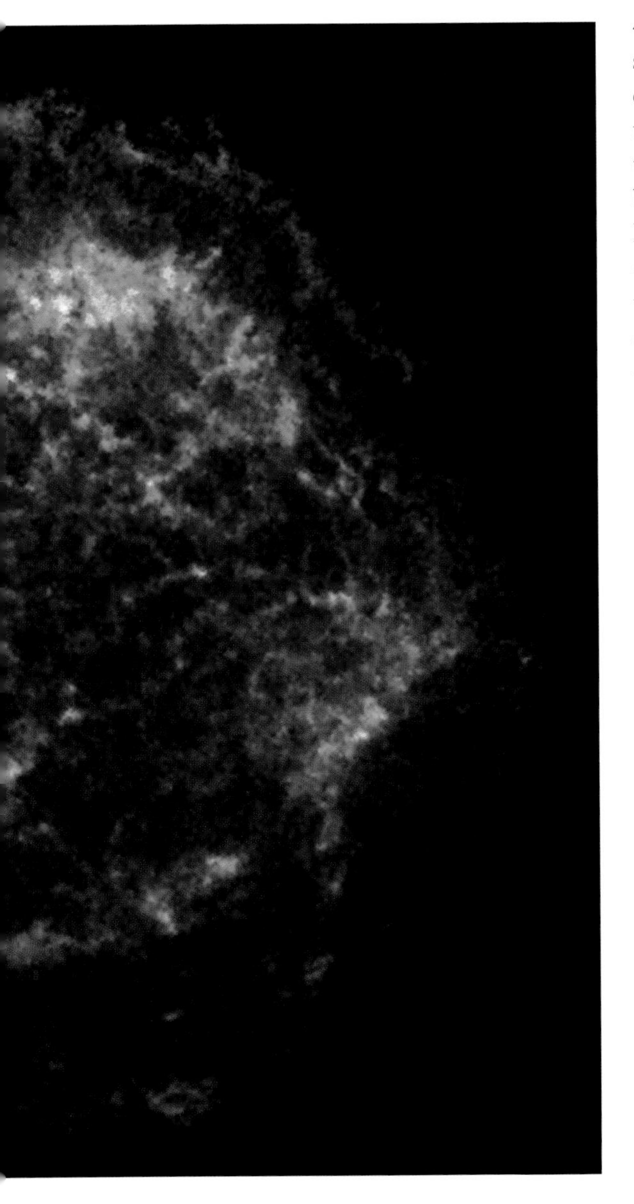

40,000 km (24,850 miles) per second. Only at this stage does the exploding star begin to brighten dramatically at visible wavelengths. Within a day or two, its optical luminosity rises to about 600 million times that of the Sun (absolute magnitude -17) – as bright as an entire small galaxy. The visible light, however, is only a tiny fraction of the total amount of energy released in the explosion, 99 per cent of which is carried off by neutrinos, most of which escape into space, travelling at, or indistinguishably close to, the speed of light.

An event of this kind is called a Type II (or 'core-collapse') supernova. The expanding cloud of debris – the supernova remnant – sweeps up and compresses the tenuous interstellar gas into which

From its rate of expansion, astronomers calculate that the explosion that produced supernova remnant Cassiopeia A occurred around 1667. The event was not reported: perhaps clouds of dust hid the supernova.

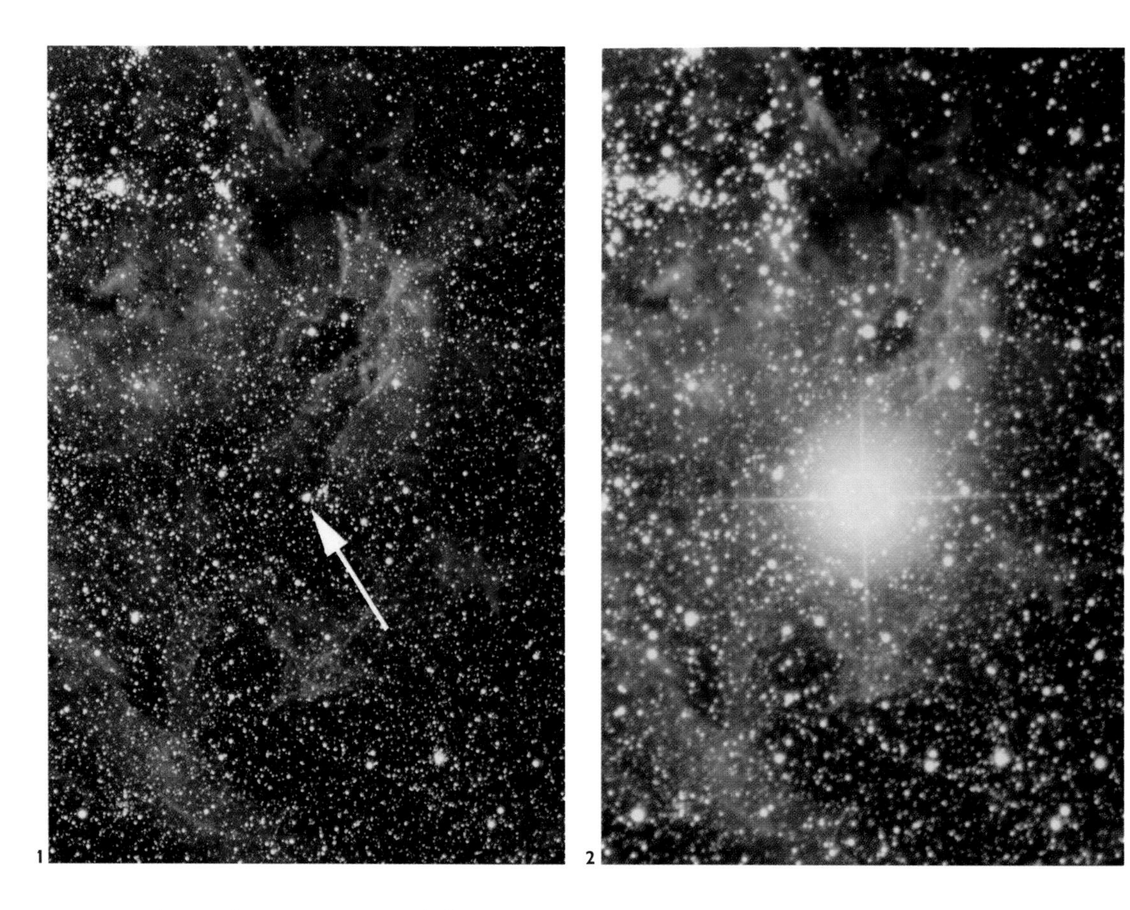

it is expanding and creates a growing hole surrounded by an expanding shell of compressed gas. Fast-moving electrons spiralling around in the trapped, compressed magnetic fields that exist in these expanding shells emit various kinds of radiation, from X-rays to radio waves.

The best-known supernova remnant is the Crab Nebula in the constellation of Taurus. Its position matches that of a naked-eye supernova, which was

1. and 2. The region around supernova 1987A before (left) and after (right) it exploded. The star that exploded is identified by the arrow.

seen in the year 1054. Since the Crab Nebula lies at a distance of around 6500 light years from Earth, the explosion that gave rise to this remnant must have happened 6500 years earlier. Another fine, though faint, example is the Vela supernova remnant. This huge filamentary cloud is about 1000 light years across, and its centre is about 1300 light years away from us. Its rate of expansion indicates that the explosion must have occurred about 11,000 years ago. At peak brilliancy, this supernova must have reached magnitude -9 and would have been 100 times more brilliant than the planet Venus.

Supernovas are rare events that occur, on average, about three times per century in a large spiral galaxy such as our own. The last one to be seen in our galaxy occurred in the year 1604, although it is possible that others may have occurred since then if they were hidden from view by dense clouds of dust.

The most recent naked-eye supernova flared up on 23 February 1987. Known as SN1987A, it took place in the Large Magellanic Cloud, a nearby galaxy 170,000 light years away. Despite its great distance, it reached a peak apparent magnitude of 2.8 and was readily visible to the naked eye. A feature of this event is that a burst of neutrinos was registered by detectors in Japan and the USA several hours before the supernova was seen by optical astronomers. This result fits in well with the theory of how Type II supernovas are formed.

THE CRAB NEBULA SUPERNOVA

The supernova of AD 1054, which created the Crab Nebula (right), was recorded as a 'guest star' by Chinese observers. It was visible in the daylight sky for 23 days and remained the brightest object in the night sky (apart from the Moon) for nearly two years. It does not seem to have been recorded by European or Middle Eastern astronomers. Some astronomers have suggested that rock carvings and paintings found in northern Arizona, USA, which show images of a circle adjacent to a crescent Moon, may represent a close conjunction of the supernova with the Moon, which took place on 5 July 1054. As the Moon moves quite quickly relative to the background stars, this event would have been visible only from the Americas.

NEUTRON STARS AND PULSARS

Neutron stars are so tiny that, although their existence was predicted as long ago as 1932, there seemed little prospect that they would ever be detected. In 1967, however, Jocelyn Bell (now Professor Bell Burnell), then a research student working at Cambridge with Professor Antony Hewish, discovered a curious source that emitted a short pulse of radio emission once every 1.33 seconds, and maintained this period with astonishing regularity. Within a few months, the Cambridge team had discovered several of these objects, which came to be known as pulsars (an abbreviation for pulsating radio sources). Since that time, more than 1000 pulsars have been detected, with periods ranging from about 4 seconds down to 1.6 milliseconds (0.0016 seconds).

Astronomers soon realized that the best explanation of pulsars is that they are rapidly rotating neutron stars, which emit narrow beams of radiation that sweep round, like the beam of a lighthouse, as the star rotates. Each time a beam points our way, we see a pulse of signal. When a rotating body is compressed, it spins faster to preserve its total amount of rotational motion (angular momentum), and when a large star collapses down to the size of a neutron star, it ends up spinning very rapidly indeed.

The magnetic field at the surface of a collapsing star grows rapidly in strength. The magnetic fields at the surfaces of neutron stars are believed to be

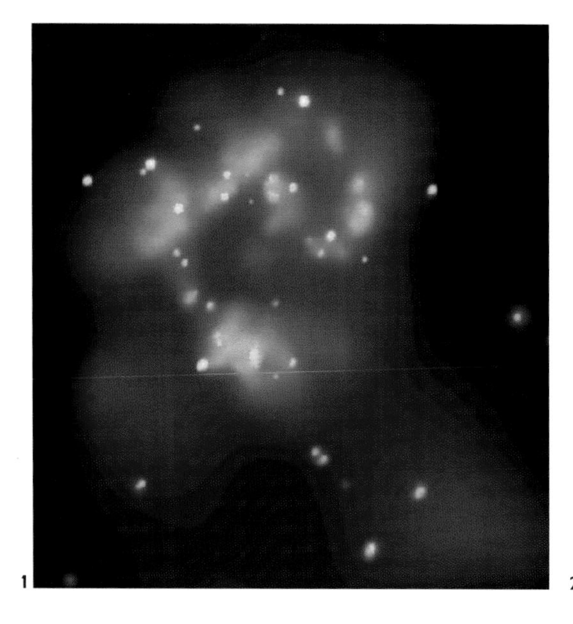
1

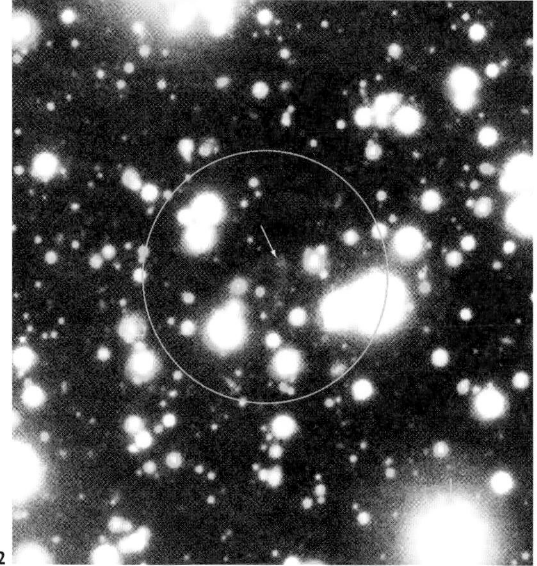
2

1. The bright points in this X-ray image of two colliding galaxies are probably neutron stars, or black holes, surrounded by gas.

2. The arrow points towards a neutron star, RX J1856.5-3754, which, while ploughing through the interstellar gas, has formed a tiny, cone-shaped nebula like the bow wave of a ship.

3. An X-ray image of the Crab Nebula, showing the central pulsar with rings and jets of high-energy particles.

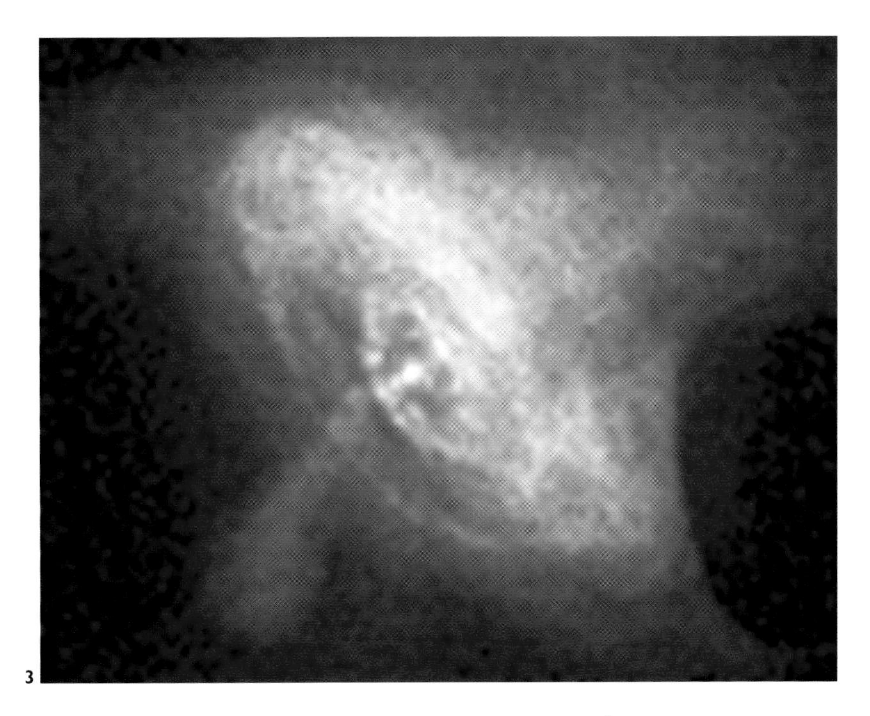
3

between 100 million and 10,000 billion times stronger than the magnetic field at the surface of the Earth. Like the field lines around a bar magnet, field lines bunch together at the north and south magnetic poles of a neutron star. Charged particles (such as electrons), accelerated by powerful magnetic forces, emit radiation that is channelled into narrow beams heading outwards from each pole. Depending on the energies of the charged particles, they can emit all kinds of radiation, from gamma rays to radio waves.

Strong confirmation of the theory that pulsars are neutron stars created in core-collapse supernovas came with the discovery, in 1968, of a

If a teaspoonful of white dwarf material were brought to the Earth, it would weigh as much as a small car. A teaspoonful of neutron star material would weigh as much as a small mountain.

pulsar, with a period of 0.033 seconds (30 pulses per second) in the heart of the Crab Nebula. The Vela supernova remnant also contains a pulsar.

Because surrounding material exerts a drag on their rotational motion, the rotation rates of neutron stars gradually slow down and their pulse periods increase. Occasionally, though, a pulsar will speed up abruptly by a tiny amount. These events, called 'glitches', are believed to be caused by 'starquakes', which occur when the outer crust of a neutron star (believed to consist of a solid crystalline layer of heavier atomic nuclei) slips, relative to the fluid interior, or slumps inwards by a microscopic amount.

Young pulsars usually have shorter periods than old ones, but there are some, called millisecond pulsars, which have very short periods (from a few thousandths of a second upwards), even though the underlying neutron stars appear to be very old. Most millisecond pulsars appear to be members of close binary systems. What is believed to have happened is that gas has flowed from the companion star on to the surface of the neutron star. Since the two stars were revolving round each other at very high speeds, the infalling material carried so much rotational motion that it caused the old and relatively slowly rotating neutron star to speed up until it was spinning round hundreds of times a second.

In some cases the intense radiation from the neutron star may eventually completely evaporate a companion star. This appears to be happening right now in the so-called Black Widow pulsar, where the companion has a mass of only 0.05 solar masses and is surrounded by a cloud of material that appears to have evaporated from its surface.

BINARY PULSARS AND GRAVITATIONAL WAVES

According to Einstein, rapidly revolving pairs of massive bodies should radiate gravitational waves – ripples in the gravitational field that travel outwards at the speed of light. As these waves are so weak, none have yet been detected directly. However, because a close binary loses energy while it is radiating gravitational waves, the two bodies will gradually spiral closer and closer together, and their orbital period will decrease. The orbital period of a binary pulsar, known as PSR 1913+16, which was discovered in 1974, and which consists of a pulsar together with another neutron star, is known to be decreasing at a rate that very closely matches what would be expected if it were emitting gravitational waves. The two stars are expected to collide in about 200 million years' time.

BLACK HOLES

Using Einstein's general theory of relativity, the German mathematician Karl Schwarzschild showed, in 1916, that if a given mass of material were compressed inside a small enough radius (now known as the Schwarzschild radius) nothing – not even light – would be able to escape from within this radius. For the Sun, the Schwarzschild radius is 3 km (1.8 miles), whereas for a 10 solar-mass star, it is 30 km (18.6 miles).

The maximum possible mass for a neutron star is believed to be between 2 and 3 solar masses. If the collapsing core of a massive star exceeds this limit, no known force can halt its collapse. Not even the immense pressure exerted by close-packed neutrons can prevent it from continuing to collapse without limit. When the collapsing star passes inside its Schwarzschild radius, its light can no longer escape, and it disappears from view. It continues to collapse until all its mass is crushed together into a point of infinite density (a singularity). All that is left is a singularity surrounded by a region of space, with a radius equal to the Schwarzschild radius, within which gravity is so powerful that not even light can escape to the outside universe. The boundary of this region is called the event horizon because no knowledge of any events that occur inside the boundary can be communicated to the outside world. The resulting object is called a black hole ('black' because it radiates nothing, and 'hole' because its powerful gravitational field can pull matter and light in through its event horizon).

Although it radiates nothing, a black hole can be detected by its effect on its surroundings. In particular, if a black hole is a member of a binary

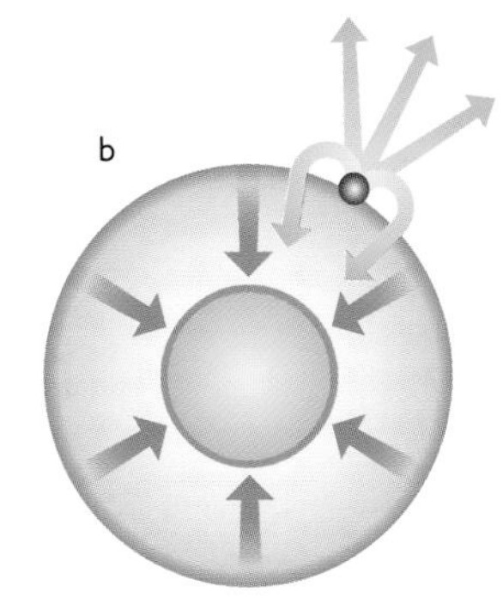

1. As a massive star collapses (a), the strength of gravity at its surface increases, and it becomes progressively more difficult (b, c) for light to escape from its surface. When the star passes inside its Schwarzschild radius (d), it becomes invisible and a black hole (e) is formed.

system, the visible star will be seen to be revolving around an invisible object. If the mass of the invisible object is more than three times the mass of the Sun, astronomers have to suspect that a black hole is involved.

If the visible star and black hole are close enough together, the tremendous gravitational attraction of the black hole will distort the star into an egg shape and gas will flow from the pointed end of the 'egg' towards the black hole. Since the star and black hole are revolving round each other at very high speeds, the stream of gas will not fall straight into the black hole but will form a rapidly rotating disc of material (an accretion disc) around it. Frictional effects, and the impact of the infalling gas, will raise the temperature of the disc to tens or hundreds of millions of degrees, causing it to emit X-rays. As hot spots in the disc revolve quickly around the black hole, the X-ray brightness will flicker very rapidly.

The best known example of a binary involving a potential black hole is Cygnus X-1, a rapidly varying source of X-rays in the constellation of Cygnus, which was discovered in 1972. It consists of a bright blue supergiant 30 times as massive as the Sun and an invisible companion with a mass

1. This blue jet is believed to be a stream of sub-atomic particles shooting out from a disc of superheated gas that surrounds a black hole. The hole, 2 billion times as massive as the Sun, is located in the core of the giant elliptical galaxy M87.

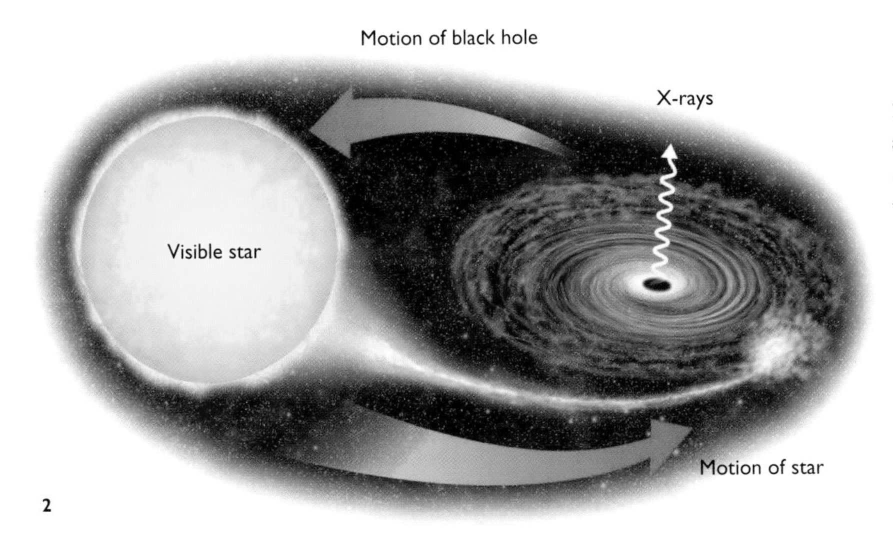

2

2. The gravitational attraction of this black hole is dragging a stream of gas from its companion star into a rapidly revolving disc (an accretion disc), which is so hot that it emits X-rays.

that may be as high as 14 solar masses. Cygnus X-1 is an example of a high-mass X-ray binary in which the visible star is more massive than the invisible companion. The error in determining the mass of the invisible object is usually less when the visible star is of relatively low mass. A particularly good example of a black hole candidate with a low-mass companion star is V404 Cygni, which consists of a visible star of about 0.7 solar masses and a dark companion of about 12 solar masses.

There are several dozen good stellar-mass black hole candidates. In the face of the growing body of evidence, there now seems very little doubt that the most massive stars in the Universe are indeed fated to end their days by collapsing to form black holes.

BLACK HOLES – THE EARLIEST IDEAS

The modern concept of black holes dates from 1916, but the idea that there might exist bodies so massive, or so dense, that light could not escape from their surfaces was first suggested in 1783 by the English astronomer John Michell and, independently, by the French mathematician Pierre Simon de Laplace (right) in 1796. Both assumed that light would be affected by gravity, just like particles of matter. They then worked out how massive a body would have to be for its escape velocity to equal the speed of light. Michell calculated that it would have to be as dense as the Sun but with a diameter 500 times larger. Both men suggested that invisible massive objects in the Universe could be detected by their gravitational effects on nearby matter.

EXPLOSIONS OF DIFFERENT KINDS

When a white dwarf has a binary companion that is a distended giant, it will drag a stream of gas from the companion. The stream of gas will flow into an accretion disc revolving round the white dwarf, and friction in the disc will cause gas to spiral downwards on to the white dwarf's surface. As more and more hydrogen accumulates on the surface of the white dwarf, its powerful gravity compresses the gas into a dense layer and pushes up its temperature. When the temperature of the layer reaches about 10 million°C, hydrogen-burning reactions switch on, escalating the temperature even further until a violent burst of nuclear energy causes the star to flare up to thousands of times its normal brilliance. Since the explosion takes place on its surface, the underlying star is unaffected. As the reaction products dissipate into space, the star fades back over the next few months to its original brightness. An event of this kind is called a nova.

X-ray bursters are sources of X-ray emission that flare up to peak brilliance in less than 1 second, remain at their peak for 10–20 seconds, then decline back to normal. These flare-ups are believed to be caused by thermonuclear explosions that occur on the surfaces of relatively elderly neutron stars in close binary systems. As their magnetic fields are weaker than those of young neutron stars, infalling matter settles fairly uniformly on their surfaces (rather than being channelled into 'hot spots' at the magnetic poles). Hydrogen burning

1. Artist's impression of gas flowing from a companion star into a pancake-shaped disc that swirls around a compact white dwarf. When gas from the disc collapses on to the white dwarf's surface, it triggers a nova explosion.

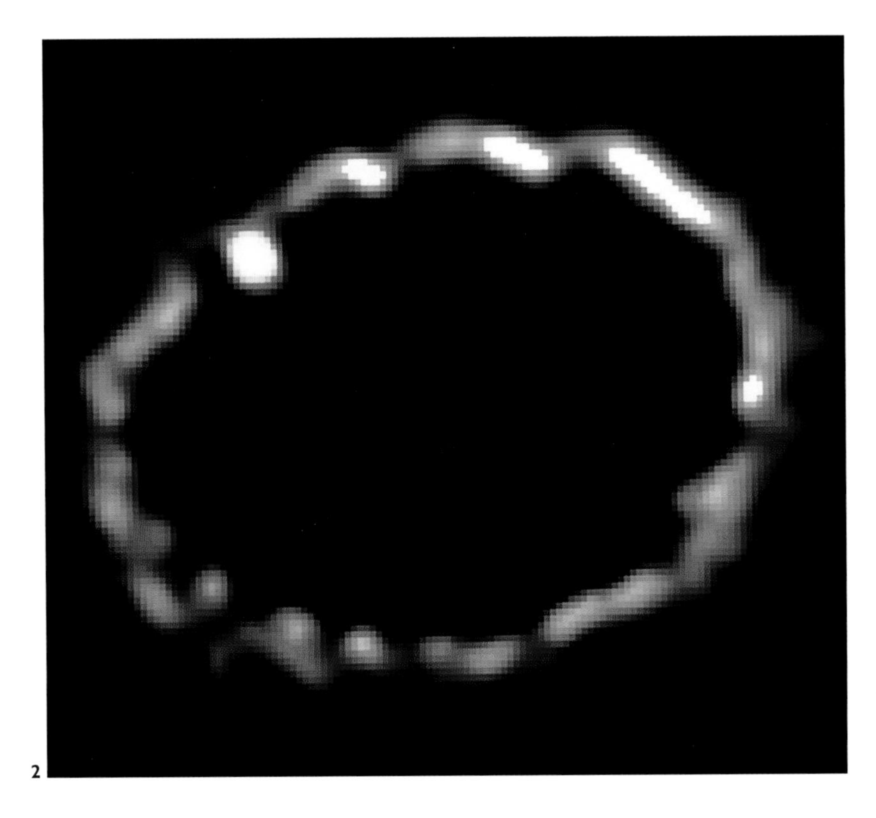

2. An image from the Hubble Space Telescope shows a glowing ring surrounding the remnants of a supernova. The ring is made up of gases ejected some 20,000 years before the central star ended its life in a supernova explosion in February 1987.

takes place on the surface of the neutron star and produces a layer of helium. By the time the layer has built up to a depth of about 1 m (3 ft), the temperature at its base has risen to about 2 billion °C. The helium then ignites in a flash, liberating in a few seconds as much energy as the Sun produces in about three days. After each burst, a fresh layer accumulates and the process repeats.

If a white dwarf is very close to its maximum permitted mass (1.4 solar masses) and is accreting hydrogen from a companion star at a gentle rate, the accumulating hydrogen burns steadily to helium, while the mass of the white dwarf continues to increase. When it reaches the Chandrasekhar limit, the white dwarf starts to collapse, suddenly heating the carbon and oxygen of which it is composed, and triggering, deep inside the star, a violent burst of nuclear energy that blows it apart. An event of this kind is a Type Ia supernova, an explosion so catastrophic that it destroys the star, leaving no central remnant. Type Ia supernovas attain absolute magnitudes of

1. and 2. The rapidly fading visible-light fireball associated with a gamma-ray burst on 23 January 1999, which, briefly, was as brilliant as 100 million billion stars. The blast took place in a remote galaxy (inside box).
Right: Close-up of the fireball and the galaxy.

For a few seconds, gamma-ray burst GRB 990123, which flared up on 23 January 1999, shone as brightly as the rest of the observable Universe put together.

between -19 and -20 (about 5–10 billion times more luminous than the Sun), and are about 10 times more brilliant than Type II supernovas.

Gamma-ray bursts

The most energetic explosions that have ever been seen are gamma-ray bursts (GRBs) – short-lived flashes of gamma radiation that can appear anywhere in the sky and which last from less than 1 second to, at most, a few minutes. On average, gamma-ray bursts are observed at a rate of about one per day. Until very recently, astronomers did not know whether GRBs were explosions occurring relatively nearby, in the outer regions of our own galaxy, or colossal explosions taking place in the fringes of the observable Universe. In 1997 an optical 'afterglow' from the fireball associated with a GRB – one which occurred on 28 February 1997 –

3. In this image, taken after the faint optical flash from a gamma-ray burst that occurred on 14 December 1997 had faded from view, the arrow indicates the galaxy, 12 billion light years distant, within which the explosion occurred.

was detected for the first time and was shown to be in a distant galaxy. Since then, several similar events have been observed in galaxies at distances of up to 12 billion light years. In order to appear as bright as they do, when viewed from such huge distances, these events have to be astoundingly energetic.

If they radiate gamma rays and visible light equally in all directions, the most luminous gamma-ray bursts shine for a few seconds as brightly as 10 million billion stars, and radiate more energy in those few seconds than the Sun will do in its entire 10 billion-year lifetime. If their radiation is concentrated into narrow beams, the energy output is less, but the total number of bursts must be far greater than the numbers actually observed (we will see a burst only if its beam is pointing in our direction). Either way, GRBs are by far the most energetic explosions ever to be seen, and ordinary supernovas pale into insignificance by comparison.

One suggestion is that GRBs are produced by hypernovas, hypothetical explosive events that are hundreds of times more luminous than ordinary supernovas. It has been suggested that a hypernova will occur if, when the core of a massive star collapses to form a neutron star or black hole, the shock wave fails to blow away the outer parts of the star. If all the star's material falls in on top of the collapsed object, the energy of the infalling matter is converted into heat and radiation, and the total output of radiation may be far greater than in a conventional supernova. No one knows, however, if hypernovas actually exist. The most popular explanation of GRBs is that they occur when two neutron stars spiral together and collide, or when a neutron star collides with a black hole. Either possibility will release huge amounts of energy, though it is not clear exactly how the energy is converted into a gamma-ray burst.

RECYCLING AND THE FUTURE

In old age and death, stars return at least some of their material to space, in the form of stellar winds, planetary nebulas and supernova remnants. Nuclear reactions that occur during supernova explosions, and afterwards in the expanding cloud of debris, produce a wide range of chemical elements. The rocks and metals that make up planets such as Earth, and many of the elements essential for the existence of life, were forged inside massive stars and spewed forth by supernovas. Material ejected by stars mixes with interstellar gas clouds, enriching them with heavier elements. The shock waves from violent events such as supernovas compress interstellar clouds, causing them to collapse and give birth to new generations of stars. Indeed, a supernova may have triggered the collapse of the cloud that gave birth to the Sun and Solar System.

About 90 per cent of the gas in our Galaxy has already been converted into stars. When all the gas is used up, star formation will cease. The last of the high-mass stars will evolve quickly, exploding as supernovas and collapsing into neutron stars or black holes. The last of the solar mass stars will survive for 10 billion years or so before turning into white dwarfs and starting their long, slow fade to the black dwarf stage. Eventually, only the dim, slow-burning, low-mass stars will continue to shine. After a few million million years, they too will run out of fuel and fade away. The era in which stars lit up the cosmos will end and galaxies will become cold, dark places, populated by the dying embers of once-brilliant stars.

1

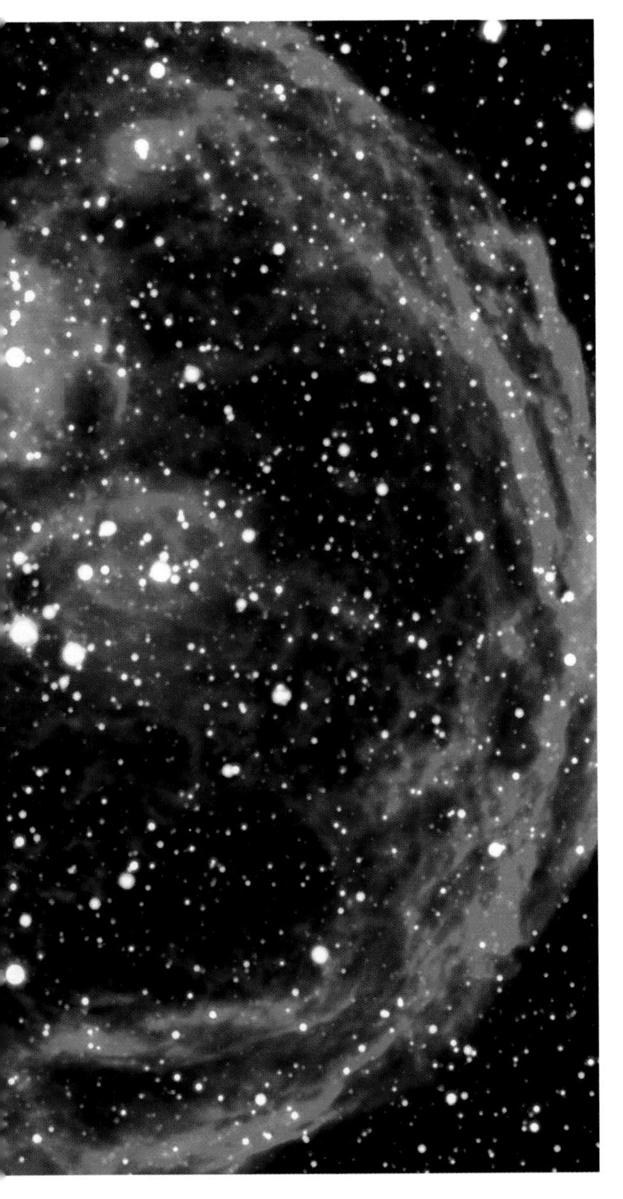

1. A 'super bubble' of interstellar gas in the Large Magellanic Cloud. The bubble, created by winds from hot stars and supernova explosions, shows how massive stars interact with interstellar clouds.

DO SUPERNOVAS POSE A THREAT?

The major threat posed by a supernova is its output of energetic gamma and X-radiation, and cosmic rays (energetic subatomic particles). At gamma-ray wavelengths, a Type Ia supernova (the most luminous kind) would outshine the Sun at a range of several thousand light years. A Type II (core-collapse) supernova would probably have to occur within a range of about 10 light years to pose a serious threat to life on Earth, but a Type Ia would be a source of concern if it were to explode within several tens of light years.

Fortunately for us, there do not seem to be any potential supernova candidates close enough to pose a major threat. The red supergiant Betelgeuse is likely to explode as a Type II supernova sometime in the future. At its distance of about 300 light years, it will reach an apparent magnitude of about -13 (several thousand times brighter than the planet Venus), but is unlikely to produce drastic effects at the surface of the Earth. However, if a major gamma-ray burst were to occur anywhere in our Galaxy, its radiation output would almost certainly have a drastic impact on life on Earth.

FURTHER INFORMATION

Books

Iain Nicolson, *Unfolding Our Universe* (Cambridge University Press, 2000). A straightforward introduction to the whole field of astronomical science.

Rudolph Kippenhahn, *100 Billion Suns* (Princeton University Press, 1993). A clear and comprehensive account of stars and their evolution.

James B. Kaler, *Stars and Their Spectra* (Cambridge University Press, 1997). A detailed account of stars, their spectra, and the H-R diagram.

James B. Kaler, *Extreme Stars* (Cambridge University Press, 2001). Describes the lives of stars and the natures of the brightest, largest, hottest, youngest and strangest stars in the Universe.

Mitchell Begelman and Martin Rees, *Gravity's Fatal Attraction* (Scientific American Library/ W.H. Freeman, 1998). A detailed account of the nature, origin and detection of black holes.

J. Craig Wheeler, *Cosmic Catastrophes* (Cambridge University Press, 2000). An exploration of the lives and deaths of stars, supernovae, gamma-ray bursts and the nature of space and time.

Patrick Moore, *Stargazing* (Cambridge University Press, 2000). A lucid guide to what can be seen in the night sky with the naked eye.

Chris Kitchin and Robert W. Forrest, *Seeing Stars* (Springer, 1997). A practical guide to viewing the night sky through small telescopes.

Monthly Magazines

Astronomy Now (Pole Star Publications Ltd, UK)

Sky and Telescope (Sky Publishing Corp, USA)

Astronomy (Kalmbach Publishing Co, USA)

Websites

NASA homepage
http://www.nasa.gov

Space Telescope Science Institute
http://oposite.stsci.edu/pubinfo

Astronomy Picture of the Day
http://antwrp.gsfc.nasa.gov/apod/astropix.html

European Southern Observatory
http://www.eso.org

National Optical Astronomy Observatories
http://www.noao.edu/outreach

INDEX

Italic type denotes illustrations

absorption lines 35-6, *35*
accretion discs 86, *87*, 88
Achernar 27
Adams, Walter S. 67
Africa 17, 22
agriculture 22
Alcor *42*
Aldebaran 18, *22*, 26
Algol 15, 18, 43
Alnilam *24*, *28*
Alnitak *24*
Alpha Centauri 27, *27*, *42*
Altair 26, *35*
Americas 17, 22
amino acids 58
Andromeda 15
Antarctic 20, 49
Antares *37*
Aquarius 22, *71*
Aquila 26
Arabs 18, 44
Arctic 20, 49
Arcturus 26
Argo Navis 17
Aries *21*, 22, 26
Aristarchus 12
astrometric binary stars 43
Auriga 26, *32*
Aurora Australis 49
Aurora Borealis 49
Australia 27
autumn 19, *19*, 20, *20*

Barnard's Star 36
Bayer, Johannes 18
Bell, Jocelyn 82
Betelgeuse 17, 18, 19, *24*, 38-41, *41*, 44, 93
Big Dipper *see* Plough
binary stars 42-3, *42*
  accretion discs 88
  astrometric binaries 43
  birth of 61
  black holes 85-7, *87*
  centre of mass 42
  eclipsing binaries *42*, 43
  gravitational waves 84
  orbital period 42
  pulsars 84
  spectroscopic binaries 43, 56-7
  visual binaries 42
black dwarfs 71
black holes *67*, *74*, *82*, 85-7, *85-7*, 91, 92
Black Widow pulsar 84
Bootes 26
brightness
  stars 31
  variable stars 43-5, *43*
brown dwarfs 62-4

calendar 23
Cancer 22
Canis Major 17, 26, *27*
Canis Minor 26
Canopus 27
Capella 26, *32*
Capricornus 22
carbon 58, 68, 71, 75, 89
Carina 17, 27
Cassiopeia 15, 26, *78-9*, 79
Castor 18, 26
celestial equator 13, 14
celestial sphere 13-14, *14*, 21
Centaurus 27
Central America 17
Cepheid variables 43, 45, 71
Cepheus 15, 43
Cetus 43
Chandrasekhar, Subrahmanyan 74
Chandrasekhar limit 74, 89
charged particles 49, 83
China 17, 81
chromosphere, Sun 47
circumpolar stars 14
Coalsack 57
colour
  stars 34, 35, *37*
  and temperature 34-5
constellations *15*
  ancient astronomy 15-19, *16*
  distances between stars 19
  identifying 24-7, *24-5*

names 15-17
zodiac *21, 22*
corona, Sun 47-8, *47, 48*
coronal mass ejections (CMEs), Sun 48-9
cosmic rays 93
Crab Nebula *76-7*, 80-1, *81, 83*, 84
Crux Australis 27, *27*
Cygnus 18, 26, 86-7

Deneb 18, 26
distances
galaxies 45
stars 31-4, *32*
Doppler effect 36, 43
double stars *see* binary stars
Draco 26
Dubhe 24-5, *25*
Dumbbell Nebula *70*
dust clouds 44-5
birth of stars 53, 59, *59*, 62, 64
dark nebulas 57
reflection nebulas *56-7*, 58

Earth 9
ancient astronomy 12
death of Sun 70
effects of Sun 49
orbit *9*, 10
seasons 19-22, *19, 20*
eclipsing binary stars *42*, 43
ecliptic 21, *21*
Egypt, ancient 21, 23, 26
Eimmart, Georg Christoph *16*
Einstein, Albert 47, 84, 85
electromagnetic waves 34, 35
electrons 70-1, 74, 75, 80, 83
elements 92
emission nebulas 53-6, *54-5*
energy
gamma-ray bursts 91
stars 65-6
Sun 47
Equator 14, 19
equator, celestial 13, 14
equinoxes *19*, 20
Eridanus 27
Eta Carinae *44-5*
Eudoxus 12

event horizon, black holes 85

flare stars 44
flares, Sun 48, 49

galaxies 92
birth of stars 59
distances 45
gamma-ray bursts *90-1*, 91
Milky Way 11, *11*
gamma-ray bursts (GRBs) 90-1, *90-1*
gamma rays 35, 83, 93
gas, accretion discs 86, *87*, 88
gas clouds *44-5, 54-5*
birth of stars 53, *53*, 59-61, 62, 92
interstellar lines 56-7
'super bubbles' *92-3*
Gemini 22, 26
giant molecular clouds (GMCs) 58, 61, 64
giant stars 43
Gorgons 15
gravitational waves 84
gravity
binary stars 43
birth of stars 59, 62
black holes 85, *85*, 86, 87, *87*
dying stars 74, 75
stars 65
Great Bear 17, 24-6
Greece, ancient 12, 15-18, 31
Greenland *20*

helium
life cycle of stars 62, 65-6, 67-8, 69, 70, 71, 75
molecular clouds 58
in the Sun 46, 47
X-ray bursters 89
Helix Nebula *71*
Herschel, William 69
Hertzsprung, Ejnar 37
Hertzsprung-Russell (H-R) diagram 37-41, *37*, 67
Hewish, Antony 82
Hipparchus 31, *31*
Horsehead Nebula *54-5*, 57

Hubble Space Telescope 64
Hulst, H.C. van de 58, *58*
Hyades *6, 22*
hydrogen
life cycles of stars 53, 58, 62, 65-7
planetary nebulas 69, 70
spectrum *35*
in the Sun 46-7, *46*
X-ray bursters 88-9
hypernovas 91

infrared radiation 35, 58, 62, 64
interstellar clouds 92, *92-3*
interstellar lines 56-7
interstellar space 53
iron 75, *75*

Jason 17
Jupiter 64
formation of 64
orbit *9*, 10
Justus van Gent *17*

Lagoon Nebula *60-1*
Laplace, Pierre Simon de 87, *87*
Large Magellanic Cloud 65, 81, *92-3*
Leavitt, Henrietta 45
Leo 17, 22
Libra *20*, 22
light
black holes 85, *85*, 87
colour 34-5
distances between stars 10
Doppler effect 36
spectrum 35-6
light year 10
luminosity
stars 34, 37-8, *37*, 41, 45, 66-7
Sun 46
supernovas 79
Lyra 26, 69, *69*

magnesium 75
magnetic fields 80, 82-3
magnitude, stars 31, 34
main sequence stars 38, 41, 62, 64, 65-6

Mars, orbit *9*, 10
Maunder minimum 48
Mayan peoples 17
Medusa 15
Merak 25, *25*
Mercury
death of Sun 70
orbit *9*, 10
Messier, Charles 56
metals 92
Michell, John 87
microwave radiation 35, 58
Middle East 15
Milky Way 11, *11*, 21
millisecond pulsars 84
Mintaka 19, *24, 28, 56-7*
Mira 43, 44
Mizar 25-6, *42*
molecular clouds 58
molecules 58
Moon 8, 9, 10, 12, 81
myths 15-17

nebulas *38-9, 40, 82*
birth of stars *50*, 53-6, *60-1*, 61
dark nebulas 57
emission nebulas 53-6, *54-5*
reflection nebulas *56-7*, 58
neon 75
Neptune, orbit *9*, 10
neutrinos 75, 79, 81
neutron stars 82-4, *82*, 88
black holes 85
formation of 74, *75*, 78, 92
hypernovas 91
magnetic field 82-3
pulsars 82, 83-4
size *67*
supernovas 78-9, *78-9*
neutrons 75, 78, 85
North Pole 13-14, *14*, 20
novas 44-5, 88, *88*
nuclear reactions 30, 52, 92
birth of stars 62
death of stars 70, 75
main sequence stars 65-7
novas 88
red giants 68
in the Sun 46-7

Omega Centauri 66
Omega Nebula *59*
Ophiuchus 22
optical double stars 42
orbits, planets *9*
Orion 15-17, 18, 19, *20*, 22, *24*, 26, *28*, *36*, 38, 56, 57
Orion molecular cloud 64
Orion Nebula *50*, 53-6, 57, 64
oxygen 68, 71, 75, 89

parallax 31-4, *32*
parsecs 33-4
particles
charged particles 49, 83
subatomic particles 48, 49, *86*, 93
penumbra, sunspots 49
Perseus 15, 18, 43
photospheres
stars 78-9
Sun 46, 47, 48
Pisces 22, 26
Pistol Star 33, *33*
planetary nebulas *68-9*, 69-70, *70*, 71, *71*, 92
planetesimals 64
planets 8
ancient astronomy 12
formation of 64
orbits *9*
Solar System 9, 10
see also individual planets
Pleiades *10*, *18*, 19, 22, *22*, 26
Plough 17, 24-6, *25*, *42*
Pluto, orbit *9*, 10
Pogson, N.R. 31
Polaris (Pole Star) 18, 25, 26
Pollux 26
precession 26
Procyon 18, 26
prominences, Sun 48, *48*, *49*
proto-planetary discs 64
protons 75
protostars 61-2, *63*
Proxima Centauri 10, 34
Ptolemy of Alexandria 17, *17*
pulsars 82-4, *83*
pulsating variable stars 43-4, 71
Puppis 17

radio waves 35, 58, 80, 82, 83
red giants 38, 41, *42*, 44, 67, *67*, 68-9, 70, 71
reflection nebulas *56-7*, 58
Regulus 18
relativity, theory of 47, 85
Rigel 17, 19, *24*
Ring Nebula 69, *69*
Russell, Henry Norris 37

Sagittarius *11*, 22, *59-61*
Sagittarius Star Cloud *34*
Saturn
formation of 64
orbit *9*, 10
Schwarzschild, Karl 85
Scorpius *11*, 17, 22, *37*
seasons 19-22, *19*, *20*
silicon 75
singularity, black holes 85
Sirius 18, *20*, 23, 26, *27*, 31, 67
Small Magellanic Cloud *53*
SOHO spacecraft *48*
Solar System, planets *9*, 10
solar wind *47*, 48, 49
solstices *19*, 20
South Pole 14, 20
Southern Cross 27, *27*, 57
spectroscopic binary stars 43, 56-7
spectrum 35-6, *35*, 56-7
sphere, celestial 13-14, *14*, 21
spiral galaxies 11
spring 19-20, *19*
star clusters *41*, *65*, *66*
'starquakes', neutron stars 84
stars
binary stars 42-3, *42*, 56-7, 61, 84, 85-7, *87*, 88
black holes 85-7, *85-7*, 92
celestial sphere 13-14, *14*
circumpolar stars 14
clusters *18*, 19
colour *34*, 35, *37*
composition 30, 36
constellations 15-19, *15*, *16*, *21*, 22, 24-6, *24-5*
distances 10, 31-4, *32*
galaxies 11
life cycles 52, 59-71, *59-61*, 74-81, *75*, 92
luminosity 34, 37-8, *37*, 41, 45, 66-7
magnitude 31, 34
mass 41, 42, 62, 66-7
names 18
neutron stars 74, *75*, 78, *78-9*, 82-4, *82*, 85, 88-9, 91, 92
precession 26
size 38
southern hemisphere 27
spectra 35-6
temperatures 36, 37-8, *37*, 41
'trails' *13*
variable stars 43-5, *43*, 71
stellar winds 61, 62, 92
subatomic particles 48, 49, *86*, 93
sulphur 75
summer 19, *19*, 20
Summer Triangle 26-7
Sun 8, 9, *9*
age of 66-7
ancient astronomy 12
celestial sphere 13
colour *65*
composition 46
death of 70
ecliptic 21, *21*
electromagnetic radiation 35
energy 47
luminosity 38, 46
position in galaxy 11
seasons 19-22, *19*, *20*
size 9-10
structure 46-9, *46-9*
sunspots *46*, 48, *49*
temperature 38, 46, 48
super bubbles, interstellar gas *92-3*
supergiant stars 38-41, *38-9*, *41*, 43, 68, 71, 75-9, 86-7
supernova remnants *72*, *76-9*, *79-81*, *81*, 92
supernovas 44, *44-5*, 61, 89, 91, 92
brightness 45, 81, 90
formation of 74, *75*, 79-81, *80-1*, 89-90
and pulsars 83-4
threat from 93

Taurus *6*, 17, 18, 19, *21*, 22, *22*, 26, 80
telescopes 30, *31*
temperature
birth of stars 59, 62
and colour 34-5
molecular clouds 58, 61
spectra 36
stars 36, 37-8, *37*, 41
Sun 46, 48
supergiant stars *75*
Thuban 26
Trapezium 56

ultraviolet radiation 35, 53, 62, 70
umbra, sunspots *49*
Universe, ancient astronomy 12
Uranus, orbit *9*, 10
Ursa Major 17, 24-6, *25*
Ursa Minor 25

variable stars 43-5, *43*, 71
Vega 26, 31
Vela 17
Vela supernova remnant *72*, 81, 84
Venus *10*
magnitude 31
orbit *9*, 10
Virgo *20*, 22
Vulpecula *70*

wavelengths, light 34, 35
white dwarfs *40*, 41, *41*, *67*, *68*, 83, 92
binary stars *42*, 88, *88*
Chandrasekhar limit 74, 75, 89
formation of 70, 71
size *67*
supernovas 89
winter 19, *19-21*, 20, 21

X-ray bursters 88
X-rays 35, *78-9*, 80, *82*, *83*, 86-7, *87*, 93

zodiac 21-2, *21*, 26